I0461382

COMMUNICATIONAL
ENERGY

HOW SHE RULES THE WORLD

VITALII SALEI

SYSTEMS THINKING · SCIENCE · EXPANDED MIND

Copyright © 2025

All rights reserved. No portion of this book, including but not limited to text, images, or any other content, may be reproduced, distributed, transmitted, or stored in any form or by any means, whether electronic, mechanical, photocopying, recording, or by any other information storage and retrieval system, without prior written permission from the author. This includes any form of publication, online or offline, in whole or in part, or in any derivative work, whether now known or hereafter invented. Unauthorized use of any material from this book is prohibited and may result in legal action.

TABLE OF CONTENTS

Introduction 1

How to work with this book 7

Chapter 1 - Rhythms of connections in the Universe..............11

Chapter 2 - Communicational Energy: The Foundation of Connections and Laws 23

Chapter 3 - The thing that connect us. The invisible logic of everyday life...37

Chapter 4 - The Communicative Energy of Consciousness. What Makes the *I* Possible?...49

Chapter 5 - Time and the hidden field of potential matter 65

Chapter 6 - Man on the path of evolution 85

Chapter 7 - Neurobiology of communication 97

Chapter 8 - Games of Consciousness and the movement of matter.. 107

Chapter 9 - Reality Code. From Zero to One. 121

Chapter 10 - Digital Alchemy (2-5): Transition from Difference to Structure 137

Chapter 11 - From 6 to 10. Algorithms leading to transcendence... 155

Chapter 12 - The Ladder of Communication Growth.......... 169

Chapter 13 - The energy of communication in the life of society 183

Chapter 14 - Language as a basis for the formation of society.. 201

Chapter 15 - Faith as a Universal Communication Protocol: God Within and Beyond 211

Chapter 16 - Digital Institutions and Communication of the Future...225

Chapter 17 - Communication Literacy......................................241

Chapter 18 - Consciousness and Beauty: perception and creation..253

Chapter 19 - Positivity as the main impulse of Being.............269

Chapter 2.0 - Contract with Life: Perpetual Motion and the Great Observer...281

Epilogue - Evolution continues..291

INTRODUCTION

It's hard to find a black cat in a dark room, especially if it's Schrödinger's cat.

We live in an era of unprecedented openness and deep perplexity. Humanity has approached new boundaries of understanding: we have decoded the genome, measured gravitational waves, and created neural networks capable of behaving almost like consciousness. But as science advances further, the question becomes increasingly clear: **who are we ourselves?** What drives us? How are we connected to this world, and why do we increasingly feel alienated in it?

The world is accelerating. More data, but less clarity from it. More choices, but also more anxiety. And with it, more loneliness, misunderstanding, internal ruptures. We know how to get to Mars, but often we don't know how to connect with a modern child. We can contact anyone on the planet, but we can't always say the right words to those around us.

We grow more connected, but to what extent do we understand their importance and depth?

Understanding the connection is not as digital but as a living thing. Not connections like Wi-Fi and file transfers, but in understanding the exchange of energy. The energy that connects atoms into molecules, people into communities and civilizations, and consciousness with itself. The energy that forms matter from simple to complex, to stable and flexible, that focuses meaning from chaos. The energy that I call **Communicational**.

This book will cover powerful communication skills, but the main focus will be on the deep power hidden behind any interaction. From the laws of physics to your personal decisions. About how **invisible connections shape visible reality**.

There will be no dogmas here. Only an attempt to build new optics. To search for a language in which you can speak with the neural network and with your intuition. To find the principle that permeates both the Universe and the gaze of a loved one.

If you feel that old explanations no longer work and are ready for thought experiments, welcome! The book invites you to take a detailed look at the main energy of the universe, Communicational energy. How it drives evolution and why it is so important that you can activate it in yourself.

Like many of you, I have probably wondered what the secret of successful people is. And with each passing year, it became clearer: **it is not about IQ or resources. Everything is decided by the ability to connect effectively.** To create, maintain and develop meaning between people, ideas, actions and oneself.

I am not a theoretical writer or an academic. And I approached the study of the effectiveness of certain communications and interactions as a businessman and practitioner.

For the last twenty years, I have been creating and developing projects in different countries, from processing plants in Cambodia and Ukraine to the production of wooden puzzles in Poland and the USA. From e-Commerce startups with a small team to agro-companies with hundreds of employees. At the same time, I was running public projects, assembling teams, looking for forms of interaction and coordination. All this was not theory, but dozens of real, numerous tests, conflicts and mastering, moments of growth and stories of failures.

My two higher educations in the humanities and my master's degree provided the basis for analysis and observation, creating the basis for the formation and design of the ideas of this book. But business and public initiatives became a testing ground for me. This is where hypotheses are tested, not in words, but in practice. Some collapsed, others remained. And more and more often, a universal principle emerged through them, what I call Communicational Energy.

In a special issue of Science magazine to mark its 125th anniversary, the editors compiled a list of the 125 greatest unsolved questions in science. The top two spots were taken by seemingly disparate but equally mysterious topics:

— what the universe is made of (95% of its mass and energy are still unknown).

— and what is the nature of consciousness (how the subjective experience of the sense of *I* arises from neural activity).

One question is about outer space. The second is about the inner world. And both involve searching for a connection that we have not yet learned to explain.

This book is not a rigorous academic work and does not have definitive truths. It is structured as a study in which intuition works with logic, and personal observations with scientific data. It results from eighteen years of study, reflection, experience and comparisons between neuroscience and cultural studies, history and biology, field physics and information theory. Here, philosophy meets science, and the personal path meets universal questions.

So, in this book, from hundreds of presented facts, you will learn how:

- Ant pheromones explain AI algorithms,
- Mirror neurons make you vulnerable to other people's moods,
- Quantum entanglement is showing up in your relationships.

The central idea of the book is Communicational Energy, and it should be clarified a little. Communication is an active process, and Communicational Energy is the basis that makes this process possible.

If we consider communication as a specific form of interaction, then Communicational Energy is the space in which such forms come into existence. It manifests itself when a stable connection arises between parts of the system, connecting and organizing the structure of visible matter.

We are living in a time of great change. Many call it difficult or pivotal. Not because there are more problems, but because the **old model of the world no longer works**, and the new one has not yet been formed.

When Gutenberg invented the printing press, it took humanity at least 80 years to realize that this was a new technology and **a new way of thinking**.

Today, we find ourselves at a similar point with global processes and cutting-edge technologies. Neural networks like GPT-4 are already conducting a dialogue, passing thinking tests, and challenging the definition of consciousness. What it is, where it comes from, and where its evolution might take us. And this is a technological challenge and a shift in the entire perception of the world.

At the same time, familiar systems are losing credibility. Politics is stuck in the past. Economics is detached from human values. Education does not prepare for the reality that will arrive in ten years. And this is not a catastrophe, but a process of transformation. Historian Arnold Toynbee called such phases the **time of creative response**: when old means no longer work, and new ones, requiring new thinking, have not yet been found.

That is why it is especially important today **to learn to see differently**. To create a new map of interaction. To find a new language and new directions of communication. And this book launches these processes.

The book was born over a long period. Some ideas matured for years, others remained questions. The path to the whole was unpredictable, as is any living evolution.

The key moment for this book came when reality itself began to confirm my hypotheses. For many years, I had been collecting all the ideas and facts into a single work, but it became a reality only when the Communicational Energy manifested itself in a new, evolutionary form.

In an era when texts, sounds and images are increasingly created automatically, it is especially important to preserve the uniqueness of human experience and the depth of individual reflection. In this sense, the participation of AI in the work on the book was like a mechanical machine grinding stone, or a calculator helping to speed up the counting. It was not a co-author, but a tool. And the main thing is not even in helping to focus, to improve the style of presentation. But that this is communication itself and the experience of this communication harmoniously allowed me to accurately reveal the ideas and hypotheses I had been nurturing for many years.

AI has become not a source of meaning here, but a means of adjustment. And, according to my observations, meaning will not appear in neural networks anytime soon, despite their ability to reproduce external form. But the fact that such communication is becoming our usual actions is only a confirmation of what is written in this book.

The main thing is this book is not generated. It has gone through all the stages of real work: from the idea and doubts to the search for form and the final issue.

Now the book is written and it is yours too. Any book is not finished until it is read. And if at least one thought from this book will sound in you, then the **Communicational Energy will manifest itself again.**

How to Work with This Book

We are the way the universe
understands itself
— Carl Sagan

Communicational Energy is not an abstraction, but an applied tool capable of changing everyday life. It can be felt, applied, and adjusted. So the book speaks not just about knowledge, but about how the quality of communication itself is transformed, as well as how new guidelines for action appear in it.

In the book, you will find:

- The main chapters, where the ideas of the book are revealed through the analysis of scientific data, personal experience and observations of processes in society.
- Cognitive blocks are short, practical exercises to turn knowledge into skills.
- Direct links to scientific sources are given at the end of the book for the convenience of systematic study.
- Glossary of terms to help you understand key concepts better.

Communicational Energy is not a magic formula. It is not an innovative approach for corporate training or a metaphor to decorate PowerPoint slides. It is a hypothesis. Based on observations, experience, mistakes, failures and growth. And subject to verification, doubt, and rethinking.

Some ideas presented in this book may seem surprising, provocative, or out of the ordinary. It's natural and healthy to challenge established ideas.

I did not try to create a universal model. Rather, my goal was to provide a holistic description of the mechanism that, in my observations, plays a decisive role in the evolution of systems, cultures, and consciousness. The intention of this book is not to convince, but to introduce a different perception. To offer the reader new reference points and a broader view of the world.

You can read in order or chaotically. But only on one condition: **do not turn reading into consumption.** Turn it into marking your own map of reality with clear routes, peaks to climb to expand horizons. And areas hidden in the depths of incomprehensible images or in dense forests of terms.

So if you find a discrepancy here, check it. Don't be afraid to argue. But before jumping to conclusions, ask yourself: **Do I really understand what I disagree with, or am I defending the old picture of the world?** Doubt has always been an important part of interaction. Scientific hypotheses, like good ideas, are not born in university departments. They survive in reality.

New hypotheses are constantly emerging in the scientific community, giving birth to hundreds of books and thousands of articles every month. Scientists argue with each other, sometimes disprove previous versions, and sometimes join forces and make breakthrough discoveries. Many hypotheses reach a dead end, but often pave the way to a new level of thinking.

But this is the power and essence of any communication: not in being right, but in discovering something new. If you find points of disagreement, then the book works as a space for understanding. End of form.

This concept is a theoretical construct and an explanation of a fundamental mechanism we observe in everyday life, although we rarely realize its significance. From my experience and observations, the decisive factor in the success of any project in life is effective communication. The ability to interact and achieve significant results through this interaction. To understand and be understood. This is where the power of Communicational Energy manifests itself.

We do not offer universal recipes. But we share tools for adjusting perception, for more conscious action and attentive living of reality.

Communicational Energy determines who survives and who disappears.

Look at the world: who is at the forefront of civilization today? Not people who extract oil or manufacture goods. But people who have proposed new forms of communication. Google, Meta, Amazon, OpenAI didn't just invent technologies, they rebuilt the ways we interact. Their capital is in the energy of communication.

From bees to whales, from bacteria to human cultures, the ability to connect has always been a decisive factor. People who can tell stories, capture attention, evoke emotion gain influence, power, capital. We live in an era of monetized communication.

So when you read this book, I hope you get:

1. Tools to distinguish between genuine communications and digital noise.

2. Examples from different fields that will expand your consciousness.

3. Workshops on setting up communication vision.

When I finished writing this book, it was the sum of my own communication experience. The experience of hundreds of disparate ideas, thoughts and phrases, united and realized in a single work.

But the creation of any book is only the beginning. The beginning of millions of subsequent communications between the author and the reader, between ideas and consciousness, between knowledge and action. Any information recorded in the text becomes the starting point of a new path. A path in which every thought is not a point of arrival, but a new bridge to understanding the hidden patterns of the world.

CHAPTER 1

RHYTHMS OF CONNECTIONS IN THE UNIVERSE

The world is not made up of things. It is made up of events and the relationships between them — Carlo Rovelli

Most people usually think that reality includes objects: a chair, a person, a star, a country. But this is rather a convenient construction of perception. Objects represent nodes in a temporary network of interactions. What seems to be solid and stable is only a stable but temporary configuration of connections. In some ways, it looks random and chaotic, but it is also natural. And it is precisely in this contradiction that the form that we call material is born.

The world does not exist as a given. It is always happening. Constantly, and at every moment, through interaction. And if you don't consider this, then you don't understand how the reality in which you live is structured.

We begin our movement into this reality to look closely at how connections arise, how they structure the world and set in motion everything that happens in it.

There is nothing truly random in the universe. What if chaos is just an order we don't yet recognize?

Even the things scattered around a teenager's room are not random. They lie according to a certain, albeit non-obvious, logic. My wife Elena has not heard of the Nobel laureate Ilya Prigozhin's idea that "in chaos there may be a path to a new order" (Order Out of Chaos). So every time we enter our youngest daughter Eva's room, a characteristic cry is heard, a signal that at one point in the Universe, in her opinion, chaos has finally defeated order.

But Prigozhin offered a different view: chaos is not the opposite of order, but another way of its existence. Order is regularity and a law regulating the connections between elements. In scientific experiments, and in everyday life, we observe many connections: from gravity to the collective behavior of ants, from quantum entanglement to two friends simultaneously sending each other the same thought in a messenger (perhaps on a Friday night).

Some connections are clear, others are still being explored. But there are also interactions that are not only not understood, but are not noticed. They seem invisible and incoherent.

Such phenomena as a cluster of stars, a heartbeat from the touch of a loved one, the movement and connection of elementary particles have nothing in common. But what if, in all these scattered objects, vibrations, impulses and collisions, there is a single principle hidden that we have not yet learned to see? All of these, we assume, are manifestations of the same phenomenon: Communicational energy.

We view Communicational Energy not as a figurative device, but as a real mechanism for forming and coordinating connections in what we call reality. This is not an additional force, but the basis of the possibility of communication. When this possibility begins to act in life, in consciousness, we notice it as communication. But by communication, here we mean not speech or language, but the ability of different elements of the world to enter into coordinated interaction.

Man did not invent communication. Even where there is no language and thought, there is already interaction. There is already a signal. We see this in living systems, in molecular processes, in the work of elementary particles. Communication does not require words. She precedes them.

Here is a simple example. A baby reaching for its mother's face does not yet know language. But his gesture is already an act of communication. It arises before knowledge, but already carries within itself a direction. This is not just a reflex, but a moment when the field of interaction becomes tangible. It's as if the connection itself is looking to form.

This gesture is no exception. Similar acts occur at many levels: between cells, in chemical reactions, in the nervous system. Gradually, the feeling arises that communication is a function of consciousness and a condition from which consciousness can grow. Maybe gravity, chemical bonds, and neurosignals are also particular forms of the same principle: the ability of parts of the world to enter into coordinated interaction.

And in this sense, communication is no longer just the transmission of messages. It is a process in which chaos takes shape and information takes direction. Maybe it is precisely the ability to create a stable connection that underlies order as such.

Communication occurs at different levels:
- Physical: interaction of particles, fields, impulses.
- Biological: molecular signals, cellular responses.
- Cognitive: thinking, memory, attention.
- Cultural: language, rituals, symbols, algorithms.
- Metaphysical: directionality as the meaning of existence.

At each level, it is different in form, but one in essence: the ability of the system to enter into attunement and structure itself through contact.

Every interaction is a form of communication. Particles exchange momentum and spin – quantum characteristics that determine how exactly they feel and react to each other at the microscopic level. Cells transmit DNA and signals. Neurons operate with electrochemical currents. Even stars interact through the solar wind in galaxies. This makes the interaction reproducible, measurable, and scientific. Everything interacts, reflects, and responds. And that means it communicates. And it is precisely in this ability to react and be noticed that the energy that we call communicational here is contained.

Today, neural networks have become one of the most vibrant communication laboratories. Artificial intelligence has integrated into everyday life almost instantly. But it's not just technology. It is a mirror in which we can see how meaning is born from many individual signals. It is an environment where one can observe how connection turns into understanding.

A study by Anthropic (2023) on the performance of AI models points out that "meaningful coherence arises from the dense relationship between representations, not simply from the size of the model." → This is the equivalent of the principle "more connection, more meaning." The higher the system's ability to achieve distributed consistency, the deeper its semantic productivity. This is not an imitation of consciousness. This is a model where communication becomes not an act, but a structure.

We see many connections. Some are clear, others are not yet. **But maybe the strangest thing about our reality isn't black holes or quantum entanglement. It's that all of this holds together in the first place.**

The paradox of the Universe is that it is infinitely mobile and unstable in essence, but does not fall apart. What maintains order in such a constantly changing system? Why doesn't it all crumble into nothing?

Physicists describe the universe with equations. But behind each of them are more than just numbers. They are an attempt to capture the structure behind which coherence is hidden. Even such open principles as the increase of disorder in closed systems (scientifically called entropy) actually speak not so much about decay as about the desire to understand why order is possible.

To record these regularities, observation, distinction, and comparison were necessary. This is the connection. Every scientific law can be the result of communication between the mind and the world.

We don't just observe the world. We read it as a text. And communication in this text is not one of the chapters. It is its alphabet.

The universe does not compose itself from objects. It weaves itself from acts of connection.

Modern science defines energy as a measure of the movement and interaction of matter, a way of transforming one state into another. Energy is stored in closed systems and this makes it the foundation of physical reality. The law of conservation of energy is evidence that a system can continue itself in time.

Communication in the classical sense is the transfer of information between subjects. But in a broader sense, we understand it as a universal mechanism for coordinating states. This makes possible the existence of language, biological forms, technologies, and even consciousness itself.

Consciousness is a dynamic process, not a static object. It arises when a complex system acquires the ability to internally coordinate. The coordinated interaction of billions of neurons gives rise to that unique phenomenon we perceive as our own *I*.

So, communication is a tool and the basis of the internal organization of any complex system. From this point of view, communication processes should be considered as:

- A space of mutual influence, not a simple line of communication

- A field for the emergence of new structures and functions
- Primary source of systemic transformations

Modern complexity theories describe how the interaction of simple elements can give rise to behavior that is impossible to predict piecemeal. Scientists call this emergence. For example:

- from billions of neurons, where each one simply transmits a signal, consciousness arises;
- from the interaction of fish, a school is formed, moving as an organic whole;
- From words and rhythm comes poetry that evokes feelings.

This is emergence: the whole is greater than the sum of the parts. Haken, H. (Synergetics: An Introduction. Springer, 1983).

But something else is more important. But something else is more important. These examples show that when a stable communication field arises, it is capable of transmitting impulses and forming supra-level structures. The connection begins to live its own life. It flows between levels—from molecule to cell, from cell to organism, and from organism to culture.

Modern science, physics, mathematics, and astronomy are the colossal work of thousands of people trying to describe reality. However, even the most accurate models remain incomplete. Scientists themselves admit that variables remain in equations without a clear interpretation, many measurements remain inaccessible to direct perception, and key phenomena are recorded only indirectly through an effect, deviation, or displacement.

We are not always able to see a phenomenon directly, but we can judge it by the trace it leaves. Just as in the dark, you can recognize a person by the timbre of their voice, the strength of their handshake, or the smell of their perfume, so in science, through interaction, form is revealed. Reality is revealed through connections.

Our hypothesis about Communicational Energy is one of many that seek to explain the structure of reality. And, as is often the case in science, it is not based on direct experimental observation. We didn't catch this energy red-handed, but we collected enough circumstantial evidence to try, like in a good detective story, to establish who started this universal mess billions of years ago.

It is based on the assumption that all known physical interactions, gravitational, electromagnetic, strong and weak nuclear are merely manifestations of one universal property: the ability to interact. Could it be this universal phenomenon is Communicational Energy? And it is the primary energy of the Universe?

When Einstein published his theory of relativity, many considered it incomprehensible and too abstract; there is a joke-legend that at first only three people understood it. When Darwin published On the Origin of Species, many called his ideas blasphemous. These stories are not about greatness, but about how new approaches initially provoke resistance, especially if they change the familiar picture of the world.

Today, we may be on the threshold of yet another shift in perception: the idea that connection may be more primary than matter. This hypothesis may seem provocative, unformed, or too broad. **But the history of science knows many cases when important discoveries began not with recognition, but with doubts and disputes.**

There is a fundamental principle in science: it is impossible to be part of a system and not bear its properties. The particle reflects the whole. Any form of the particular carries within itself the laws of the entire structure. This is why disciplines such as synergetics (Haken, 1983), the science of how chaos begins to self-organize when a connection is created between elements, seek to find universal patterns of interactions, including communication.

So if our hypothesis is correct, and connection is primary to matter, the next logical step is to look inside the system itself. Inside ourselves. A part always carries the properties of the whole. We are inhabitants of the Universe and are woven into its logic. We are made according to the same laws as it is. And the Christian term about man made in the image and likeness takes on a new meaning. We repeat the Universe not in form, but in function. Every time we choose, comprehend, love and achieve, we do the same thing it does: we connect and unite.

It is necessary to emphasize the special status of humanity in this system. We are an element of the universe and a unique case of a system that has acquired the ability to self-knowledge. Modern neurobiological research (Edelman & Tononi, 2000) shows consciousness arises as the product of coordinated interactions between billions of neurons. The ability to understand others, create, and think in abstract categories reflects the amazing complexity of systemic relationships within the neural structures of the brain.

And if all the processes of our mental activity are carried out through communication mechanisms, a reasonable assumption arises that the Universe itself may be structured according to similar principles.

---

*We repeat the Universe in every decision
we make, in every functioning we do.*

We think we understand the structure of the world. But this understanding is always projective. The deeper we look, the more complex the structure itself becomes. Science is once again confronted with the image of communication not as transmission, but as a complex configuration in which everything influences everything.

The world is not linear. It is multi-layered and multi-dimensional. Modern physics speaks of at least four dimensions. String theory suggests at least ten dimensions.

String theory, one of the main hypotheses of modern theoretical physics, states that elementary particles are not points but tiny vibrating strings. Their vibrations create the properties of matter. For the theory to work, 10, 11, or even 26 dimensions are needed. This is still a hypothesis, but it provides a powerful image of the Universe as a system of mutual oscillations and connections. Greene, B. (1999). The Elegant Universe. W.W. Norton.

Even the most sophisticated models of the world leave the main thing behind the scenes: why all this works together at all. Physics describes forces, fields, and particles, but it cannot describe the principle of consistency itself. That invisible mechanism, thanks to which all this does not fall apart instantly.

And here's what's interesting. **Physicists have been searching for so-called Dark Energy for over 70 years.** A mysterious substance that, according to cosmological observations (such as the redshift effect and the Planck telescope), makes up about 68% of all the energy in the universe. No one has seen it. But it seems to affect everything: it accelerates the expansion of space, it affects matter, but it is not directly detectable.

Some interpretations of quantum physics and information theory suggest that dark energy may be related to information processes at a fundamental level (Padmanabhan, Emergent Gravity and Dark Energy, 2005).

Dark energy leaves no direct evidence. But it creates effects that change everything. It is not magic, just a different scale of visibility. Maybe we look at the universe like a fish looks at the ocean: it feels the pressure, but it does not know that it is water.

And here the question is appropriate. What if this energy is already in our hands? What if this is the energy of Consciousness? The Energy of Connection?

From the point of view of the Universe, man is insignificant: one of 8 billion forms of organic carbon, living on a tiny planet at the edge of the galaxy. But man is unique. He is the only one who can know the fact of his participation in the existence of the Universe. And here, Consciousness manifests itself not as a side effect of biology. Maybe it is the instrument by which the Universe itself observes itself.

If the communicational energy hypothesis is correct, it changes everything

---

In these chapters, we will look at how this energy manifests itself in quantum physics, biology, evolution, consciousness and culture. But most important, we will try to understand: can communication be a condition of life and its essence. Can it be not a consequence, but a cause.

This book is not just about the world around us. It is about a view and a way of seeing this world. We will look differently: at time, at consciousness, at interaction, at man himself. Because ultimately it will not only be about science or philosophy. It will be about you. About how you are structured.

And why you can change reality.

Chapter 1 Summary:

- Communication is not a human invention, but the primary mechanism of coordination that links everything from particles to culture.

- Energy is movement and a way of holding and transforming forms.

- Communicational energy is a hypothesis that unites matter, consciousness and information through the principle of communication.

- The connection is not always visible, but it is this that allows chaos to become order, and the particular to be part of the whole.

- It may be this energy that is the fundamental mechanism of reality.

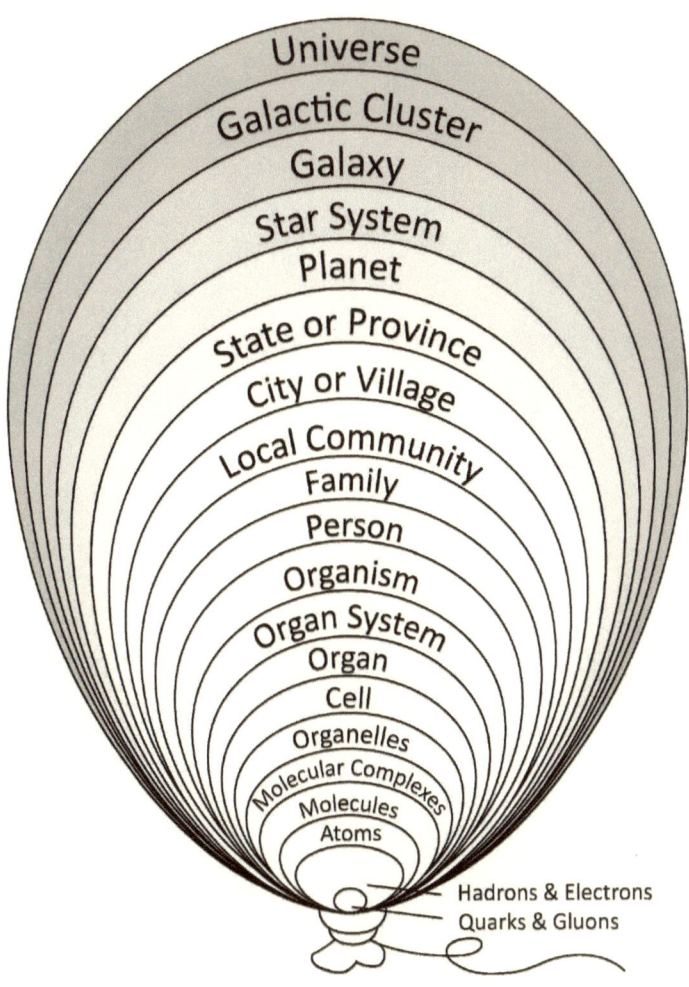

Communication levels of reality

CHAPTER 2

COMMUNICATIONAL ENERGY: THE FOUNDATION OF CONNECTIONS AND LAWS

*If you want to understand the universe,
think in terms of energy, frequency and
vibration — Nikola Tesla*

Before we move on, it is worth repeating our starting point. The first chapter dealt with communication as a universal mechanism of connection. Now it is important to distinguish: communicational energy and communication are not exactly the same thing.

We call communication everything that has manifested as an act of connection: speech, gesture, signal, interaction. Communicational energy remains in a pre-manifest state, creating the possibility for the emergence of connections. A field that does not transmit signals, but prepares the basis, organizes logic, creates a framework within which signals, meanings, forms can arise.

In scientific language, field is not a space, but a distributed entity that determines the behavior of particles and processes at each point. Gravitational, electromagnetic, quantum fields, all do not send signals directly, but structure the possibility of these signals. The field is not associated with a specific point; it is not fixed as an event. It is the condition in which events become possible. This is how Communicational Energy works. It does not send a message, does not interact directly, but prepares the possibility of communication.

Everyone knows the admiration of blooming fields of lavender or sunflowers. Everyone has walked through spring gardens where cherry or apple trees open up in a white and pink glow.

Imagine that our communication is plants. They grow, blossom, and bear fruit. And the fertile soil where all this grows, the place where all this can blossom and bear fruit – this is the field of Communicational Energy.

Maybe once we were in a single field of communication, like Adam and Eve in the Garden of Eden, where everything was already agreed upon. But then we left. We took with us the fruits and seeds we had tasted. And since then, everyone has been cultivating their own communication soil independently.

Think about it, are you a good agronomist? What do you know about the environment in which your connections grow? What kind of tree do you want to grow in this world, and from what seeds?

It is important to clarify our position on scientificity. None of the fundamental laws of physics have been proven in a strict mathematical sense. Newton's laws, Maxwell's equations, Einstein's theory of gravity are not absolute truths, but effective hypotheses accepted because they produce predictable and useful results and allow action.

Science is not a temple of truth, but a marketplace of hypotheses. Every law is a temporary contract with reality, valid until the first refutation. Only philosophers can afford the luxury of searching for truth, while the scientific world uses practicality as its main currency.

As Niels Bohr said, "Predictability is the main criterion of truth in science." So modern science increasingly considers physical law not as truth, but as a working contract. It remains if it produces results. And when a new model appears that explains more and more accurately, the old one gives way to the new one. This is not a weakness of science, but its strength. It does not cling to dogmas; it improves its language of communication with reality.

The most striking example of science going beyond common sense is quantum physics. It explores the invisible, the insensible, the incredible. Particles that behave like waves. Waves that disappear when observed. And even interactions that are instantly transmitted between distant particles, contrary to the idea of distance.

Many renowned scientists have made no secret of the fact that quantum mechanics is like magic.

Richard Feynman, a Nobel Prize winner, admitted: "I think it's safe to say that nobody understands quantum mechanics."

And when discussing the fine structure constant, a mysterious dimensionless number that determines the strength of electromagnetic interaction, he added: "A magic number that comes to us without any explanation... is the most amazing thing in nature."

Behind dozens of theories and experiments and hundreds of formulas, something more is emerging: as if we are touching the deep architecture of the world, which we cannot yet explain or grasp. That is why quantum physics inspires not only scientists but also philosophers, artists, and engineers. It works, and at the same time, it requires a new language to be understood.

We do not insist that the Communicational Energy hypothesis expresses the final truth. We say this hypothesis gives us an economical and productive way to describe the connections in reality, including both physical interactions and the processes of perception, consciousness, and cultural evolution. And if the hypothesis shows patterns, it deserves a place in the scientific dialogue.

Adolf Grünbaum, an influential philosopher and rigorous methodologist of science, once said: "It is not metaphysics that spoils science, but metaphysics that is not limited by evidence."

He did not reject metaphysics, he demanded discipline from it. In the classical sense, metaphysics is the reflection on the foundations of reality that are not reducible to observable phenomena.

We continue to move forward, and this chapter is not a step away from science, but an attempt to collect evidence from different fields of knowledge. We will delve into the manifestations of communicational energy in physics, biology, and neuroscience.

Communication, as we have specified, can be recorded and described. Communicational energy is hidden, but it structures reality, which can become visible, distinguishable, and meaningful. So the correct proof of our theory will be the analysis of traces of this energy in the physical processes themselves where there is stability, repeatability, and attunement.

And these manifestations can be seen on all scales: from galaxies moving away from each other according to the laws of cosmological expansion (73.4 km/s per megaparsec according to the Hubble Space Telescope), to the quantum level with its dynamics in superpositions and instantaneous coherence of entangled particles. Physics says everything that is stable, everything that holds in time, is always exchanging something. Energy. Momentum. Information.

But exchange is impossible without a stable connection. And this connection is communication.

The four known physical forces, gravity, electromagnetic, the strong and the weak, hold particles and objects together. They connect but do not explain how stable structure, difference, and direction arise. Maybe what gives these forces their architecture is what we call Communicational Energy.

Physical forces hold matter. Communicational Energy holds the meaning of interaction, its direction, distinguishability, and structure. It is what forms boundaries, makes contact possible, sets the mode of difference. Without it, neither a particle will find a partner, nor an addressee signal. It is what determines what and how arises in this world.

If Communicational Energy determines the structure of interactions, its manifestations should be noticeable not only at the level of physical processes but also in living systems. Where there is no external control center, but the elements still gather into stable configurations. This can be seen in insects, in microbial colonies, in neural networks.

For example, ants build bridges from their bodies. They have no plan, no coordination from above. But through local signals and simple reactions, they form a workable structure. Each action taken separately is primitive. But taken together, a whole emerges. This is no exception.

We will return to biological examples in these chapters. Such observations give reason to speak of Communicational Energy as a systemic mechanism, and not as a particular effect. Maybe communication is a series of events and a field in which the elements of the system can tune in to each other. This field does not transmit content. It creates the possibility of coherence.

On this basis, a working definition can be given:
Communicational Energy is a directed exchange of states between elements, leading to stable interaction and the emergence of structure.

But if we are talking about a scientific model, how can we measure this energy?

What metrics can record its presence? There must be signs that distinguish a system where it manifests itself from a system where it does not. How can we distinguish a living, coordinated system from a random set of elements?

Communicational energy reveals itself through three indicators:

- topology of connections - how the network is structured, who is connected to whom, and what is the density of interactions,
- stability - the ability of a system to maintain its shape in the face of external and internal signals,
- coordination ability - how quickly and accurately its parts come to act together.

The higher these parameters, the more Communicational Energy in the system. And therefore, the higher its potential for learning, evolution, culture, and thinking.

Connection is not a bridge between objects. It is what the objects themselves emerge from.

Modern physics shows that fundamental particles do not exist as isolated entities. They exist in a superposition of states, entering into instantaneous correlations no matter the distance, this phenomenon is known as quantum entanglement.

Alain Aspect's experiments (Aspect, 1981–1982) confirmed this at the level of laboratory observations. And in 2017, the Chinese satellite Micius showed the preservation of quantum correlation between photons at a distance of over 1,200 km (Yin et al., 2017).

This is not about signal transmission, but about the emergence of coherence, without relying on space and time. This may show the existence of a deep level of communication that cannot be reduced to mechanical interaction. All this speaks for physical reality being based not on matter, but on information. If information is primary, then we can assume that communication is the mechanism for the creation of form.

Physicist John Wheeler proposed the formula It from bit: everything physical arises from informational differences, from answers to questions (Wheeler, 1990). In this logic, matter is not the basis, but the consequence. The primary is the structure of distinction, the logic of connection, the code. And if this is so, then at the source is not a particle, but an act of communication.

This line is also supported by modern interpretations:

- Tegmark (2017): "Life is not just matter. It is an information-processing process. Information is what structures movement, creates stability and form. Without it, the universe would be a disordered plasma." (Life 3.0)
- Vanchurin (2020): "If we imagine the Universe as a neural network, the information flowing in its structure is the energy from which form, logic, and time are born." (The World as a Neural Network)

These approaches allow us to consider communicational energy as a force that makes possible coherence at the level of the information field.

But if information is at the core of reality, the next question is: what happens to it under conditions of uncertainty, especially where there is no clear order?

One way to understand this is through the concept of entropy. In physics, it is a measure of chaos and uncertainty: the higher the entropy, the less predictable the system.

In information theory (Shannon, 1948), entropy describes the data needed to accurately describe the state of a system. Each message is a contribution: it either increases or decreases entropy. Communication in this context is an exchange of signals and a way to **manage chaos**. It reduces uncertainty, limits the set of options, and structures choice.

Erwin Schrödinger wrote about this in his book *What is Life?* (Schrödinger, 1944) that organisms survive by absorbing negative entropy. Consciousness, as the highest form of life, also minimizes uncertainty by selecting and structuring information.

In this sense, consciousness is an anti-entropic form of communicational energy: not just a reaction, but the ability to maintain internal coherence in the face of external chaos.

Complex systems – biological, social, technological – benefit precisely to the extent that they can restrain the growth of entropy. Their stability and ability to develop depend on the level of communication consistency.

Modern astrophysics adds another important detail. Stephen Hawking has suggested that information entering a black hole does not disappear, but is preserved at its event horizon (Hawking, 2005). This means that even where matter disappears, information can be preserved.

If information is preserved even under the most extreme conditions, then we may be dealing not just with a physical phenomenon, but with a field that maintains the structure of distinguishability, regardless of its material carrier.

Information behaves not like matter but like light: it does not disappear, even when it seems to. It may be precisely this ability to retain difference—even where form disappears—that points to a deep architecture of the universe in which meaning is more important than matter.

Scientists continue to search for the nature of dark energy and dark matter, two quantities that, according to cosmological models, make up about 95% of all mass-energy in the Universe (Planck Collaboration, 2018). At the same time, attempts are being made to recreate antimatter, the same one that should have arisen in equal quantities with matter after the Big Bang.

According to the standard model, matter and antimatter should have balanced each other out — but they didn't. Matter remained, antimatter disappeared. Why?

Antimatter is the symmetrical opposite of matter: its protons and electrons are mirrored in charge, but identical in mass. What could be the opposite of existence, if not non-existence? If annihilation occurred, complete equilibration, then there was a moment of connection. A contact between two poles: presence and absence.

What can be the carrier of such a connection? Only that which allows both states simultaneously, both presence and disappearance. This allows us to assume that on the border of the existing and the rushing, we encounter another type of field. Not matter and not antimatter, but a field of correlation. Let us call it a hidden form of communicational energy, acting on the border between being and its absence.

Physicists claim that almost all of the Universe is a dark substance we cannot yet directly observe. We suggest that it may not be a substance, but a field of coordination. A field in which new levels of existence are formed.

And visible matter is only a local crystallization of this energy. The trace is the point where the structure manifested itself.

We are used to thinking about matter as something dense, tangible, having a form, mass, boundaries. But perhaps everything that exists is simply something that was connected. That which entered into a stable interaction became distinguishable and therefore received a chance to exist.

In this optics, the Universe is not so much a physical system as a system of transmission and retention of information. Not matter as such, but communication is a condition for its appearance.

Communication is the act of turning the possible into the real.

Inside each proton, one of the basic building blocks of matter, are three quarks. But if you look deeper, you discover something surprising: these quarks make up only about 2% of the proton's mass. The rest of the mass comes from something else entirely. The gluon cloud. The binding energy between quarks. Gluons are elementary particles that act like glue and bind quarks together.

Modern quantum chromodynamics (QCD) shows that the gluon field around the quarks not only holds them together, but also has mass itself. This means that the binding energy is transformed into substance.

Gluon clouds do not hold quarks together; they self-interact, creating a stable structure from which everything we consider *things* is born. What was considered an invisible force turned out to be a material medium. The object does not create the connection. The connection creates the object.

Imagine you bought a heavy watermelon of ten kilograms. You brought it home, cut it, and inside there was only 200 grams of juicy pulp. The rest was emptiness. Air. As if you bought a shell. You would be surprised. Maybe even feel cheated.

But this is exactly how all the matter that makes up your world is structured. You are literally made of the weight of interactions, not of substance, but of connections. Physics says: 98% of your mass is not substance, but gluons. The paradox is that you literally do not exist without interactions. You are not a sum of particles, but a by-product of containment. If gluons stopped connecting, you, philosophy, biology and this whole book about connections would disappear.

Physicists suggest that gluon fields can exist without quarks — as so-called glueballs — pure clots of gluon energy. They have not yet been directly detected, but mathematical models confirm their plausibility. This means that binding energy can exist not just inside matter, but also beyond it.

There is another important scientific example: the Higgs field. It is essentially an invisible medium that uniformly permeates the universe. Particles, moving through it, experience resistance, and this interaction gives them mass (Higgs, 1964). Not because something is embedded in them. But because they interact with the field. It is this interaction that makes them different, measurable, and real.

We are used to thinking that things exist, and then they interact. But maybe it's the other way around: things exist because they interact. It's like the constellations: their lines aren't drawn in the sky but we see a structure in them because there's already a connection in our heads. Physics is gradually catching up with this realization: it's not mass that creates form, but coordination.

This may be the universal mechanism: the form does not create the interaction, but the interaction that makes the form. The world becomes visible when it passes through the field of agreement.

The same principle applies to biology. In the early stages of life, amino acids didn't just collide, they responded to each other. They formed stable bonds, from which proteins emerged. Then, cells, organisms, microbiomes. And at each stage, development occurred not because more matter appeared, but because new levels of connection arose.

Communication gives birth to structure. Structure gives birth to consciousness. And consciousness is already a system capable of giving birth to new forms: languages, cultures, technologies, scientific models.

Thought is a special form of material manifestation, distinguished by a more subtle organization. In the human brain, 86 billion neurons form a dynamic network, where each element can interact with thousands of others.

The nature of consciousness is revealed not as a static object, but as a product of internal coordination and complex information processing. The human being exists as a unique intersection of countless flows, from biological processes to digital interactions.

The physical component of a person is a material form that has arisen from complex interactions. The mental sphere functions as a holistic ecosystem of meanings that exists thanks to the constant process of communication.

We are already living in an era when communicational energy is no longer a function of a separate organism. It goes beyond the body, becoming part of the technosphere: built into neural networks, algorithms, and distributed intelligent systems. And we see that it has always been more than just an awareness of the world. It is an eternal tool for its formation. Even when external conditions change and transform.

Communicational Energy is not one of the forces, but a condition under which any forces can manifest themselves. The Universe is not a set of objects, but a field of agreements. We exist because a communication mechanism works between us and within ourselves.

So, it all starts not with matter. It all starts with communication.

Chapter 2 Summary:

- Communication is not a side process, but a universal law of interaction.
- Information is primary. At the level of quantum physics and cosmology, it determines the behavior of matter.
- Entropy is a measure of chaos, and communication is an anti-entropic force: it brings order, reduces uncertainty, and creates stability.
- Consciousness is a form of anti-entropic communication, a structure that minimizes uncertainty.
- Connection is primary: it is not objects that create relationships, but relationships that shape objects.

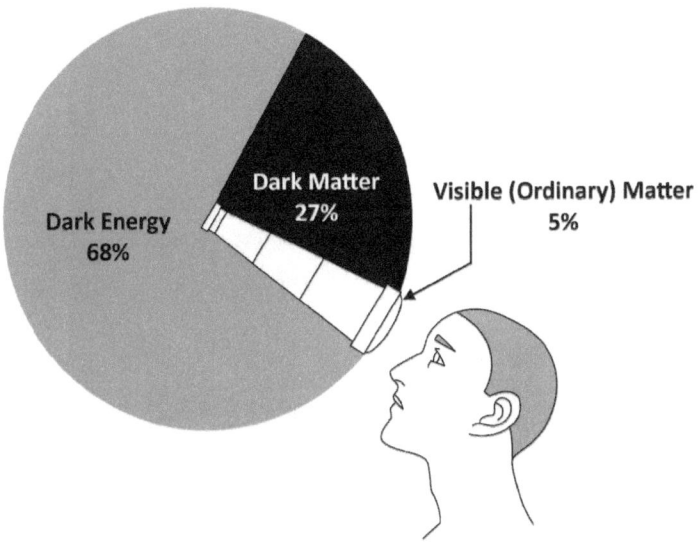

Composition of the Universe (by Energy & Matter)

Chapter 3

The thing that connect us. The invisible logic of everyday life

The only thing that gives meaning to life
is human connections
— A. de Saint-Exupéry

Communicational Energy is not just a philosophical model or a scientific hypothesis. It is a living practice, woven into our every day. Not something distant and abstract, but something that manifests itself in every gesture, look, and question.

In the book, we develop a hypothesis about the Universe. But we also want to show that understanding communication can change lives.

This chapter begins the transition from philosophy to practice. We will examine:

- how to notice hidden connections,
- how to build communication correctly
- how to shape reality through understanding the processes around

That is why cognitive blocks appear in the book. These are small elements that will work as exercises for the consciousness. Their goal is for knowledge to become skills, and skills to lead to a new quality of life.

Maybe your day begins like most modern people, not at dawn, but with a signal?

The fluctuating vibrations of the alarm clock start our new day. More precisely, not the alarm clock, but the smartphone. And to be precise, your «alarm clock-phone-photo-notebook-navigator».

The screen is the first contact of a new day. You haven't even opened your eyes yet, but you are already communicating with reality.

Scientifically, it all starts with a sensory-informational cascade: one touch and the visual cortex, the limbic system, the prefrontal areas associated with attention and decision-making are activated (Foxe et al., 2002). A simple gesture is already an interaction. Already the beginning of a connection.

We usually think of morning as simply switching on. But it may be a moment of alignment with a distributed system of norms. The world no longer opens—it loads.

Astrophysicists assume that the entire Universe originated from a single point. This is difficult to imagine and impossible to verify. But here is what can be verified easily: **man has already concentrated almost his entire world into a material point the size of a palm.**

And everything you experience during the day – thoughts, routes, signals, conversations – is already entered into this small smartphone. Communicational Energy is not hidden somewhere in theory. It is already in the palm of your hand.

Human life takes place within a complex structure of interactions. We are in constant exchange: thoughts, words, glances, icons in messengers, routes of movement, and even silence. This is an unnoticeable but continuous process. It organizes our everyday life, determines the rhythm, fills it with meaning and — paradoxically — often remains unnoticed.

We say: "the connection is lost" and we mean not only the cellular signal. The feeling of being included disappears. This leads to anxiety and confusion. Connection is a technical term and a basis of the feeling of life.

You may have noticed: the book does not move in a strictly linear fashion, like a scientific dissertation, but in waves — moving from quantum physics to everyday gestures, from matter fields to instant messengers and Mondays. This is not a structural liberty, but a reflection of the nature of Communicational Energy: it does not belong to any one area of knowledge. It penetrates everything — from neurons to markets, from elementary particles to smartphone screens.

Modern science increasingly confirms: complex systems, biological, cognitive, social, do not move along straight trajectories. They develop as networks, through impulses, instabilities and modulations. That is why we study chapters and waves of a single field. That is why you are holding in your hands not a textbook, but a conversation, a living one, moving from spectrum to spectrum.

Sometimes, to understand how the world works, you don't need to read anything. You just need to exhale. And notice how in the morning, almost without looking, you reach for the switch. This gesture has the same logic as the launch of the Universe: expectation, impulse, connection, activation.

Even the simplest things—like turning on the light, choosing clothes, leaving the house—are not autonomous. They are embedded in a complex system of signs, memory, rhythm, and anticipation.

From the point of view of semiotics and anthropology, everyday actions are not a set of instincts. They result from our integration into sign systems.

Take Monday, for example. It is a day of the week, but it is an agreed-upon arrangement, a symbol of order, written into billions of diaries, consciousnesses, and planners.

Edward Hall in The Dance of Life (1983) showed that time perception is not just a biological function. It is a communication structure embedded in culture. Days of the week, schedules, social rituals are hidden maps of interaction we rarely realize, but follow.

We don't just move around the city. We follow routes made up of earlier decisions, expectations, internal tempo. Even the trajectory in the supermarket is a dialogue with the system. Every nod, every turn of the head, every "uh-huh" is a response to the rhythm of the world. A result of previous connections. A premonition of new ones.

We live in an environment built of words, numbers, and signs. Communication is not an addition to reality. It makes reality possible and tangible. It turns the random into meaningful. The disorderly into a route.

Every day we ask dozens of questions: how to get there, what to buy, when to call. We literally live from request to answer, or at least to its expectation. We delegate more and more to digital assistants, but we ask the same ancient question: "Who are you?", "Are you here?", "Can you hear me?"

Modern neuroscience confirms that the mind operates on the principle of continuous inquiry and prediction (Friston, 2010). The question is a sign of ignorance and an uncertainty-reducing mechanism that activates the dopamine system in anticipation of a connection.

Even the search bar is a new shell of the ancient mechanism of meaning formation. We are constantly searching. And if you look closely at this process, it becomes clear: life is not just actions. It is a continuous communication with reality.

Everyday life is a network of micro-meanings united by rhythm.

But there is another level. The level of questions without an addressee. They come suddenly, most often in pauses. Not when we speak, but when we fall silent: Who am I? Where am I going, or even running? What drives me? What is the meaning?

These questions do not give direct answers. They do not add benefit, but they strengthen presence. They do not specify the route, but they lift us above the map. They work like mirrors, they truly show the one who asks the question for the first time. And here something amazing occurs. The existence of a question is already a form of connection. Between parts of consciousness. Between personality and time. Between a person and the world. The question does not require an answer. It requires awareness that the connection exists.

What we call the meaning of life is the connecting energy between the individual and the world. It is not received ready-made, it is created: as a form, a process, and a connection.

As a form, it provides stability: we understand who we are and where we come from. As a process, it provides direction: we move, overcome, and grow. As a connection, it fills life: we feel we are not alone.

Meaning is not a material object that can be touched, proven or inherited. It manifests itself in the tension between reality and its possible continuation. It is not a static object, but a dynamic interaction.

Meaning does not arise simply from existence, but from integration, connection, and generation. It is not always expressed in a clear answer – sometimes it is felt as a direction.

Now let's put the question aside.

Have you ever thought about why the meaning is initially hidden from us? Why don't they give us a personal map of meanings along with our medical records at the maternity hospital? Or at least once a year, as a system update, wouldn't we receive a message: "Dear user, your meaning firmware is out of date. Please update it to work correctly with current communications."

Well, okay, personal update is expensive. But you could at least hang a banner on the Moon. In large letters. So everyone would understand it right away.

But no. The meaning still needs to be sought. Again and again. And there may be reasons for this. I have found three.

First, that we have become conscious beings, searching for meaning, is an absolute universal randomness and absurdity. Second, there is no meaning because it is not built into the structure of existence, and everything is based on improvisation. Third, one of the features of meaning is that it is initially hidden. It cannot be given out, it must be collected. Constructed. Lived.

We don't know yet which version is true. But in these chapters, we will try to figure it out. Maybe the path to meaning is its main form.

To live meaningfully means to be in harmony with the structure of the Universe: to experience freedom as a form, to live realization as a process, and to experience love as a connection. From these three threads, what we call meaning is woven.

It's not about purpose. It's about inclusion. **Meaning is the energy that says: you're not an accident. You're part of a pattern.**

Carl Jung wrote the deep questions "Who am I?", "What is meaning?" are not rational tasks. They are born from the collective unconscious, an ancient field of memory that connects the personal with the eternal. When we think about ourselves, we come into contact with all of human history. When we listen to these questions, we come into contact with something much larger than the individual *I*. With something that was before us and will be after us.

Modern culture is designed to distract us from these questions. Constant busyness, multitasking, streams of news and worries. We become masters of external communication, but lose sensitivity to internal communication. We talk more and more, but understand less and less.

Research shows that when we constantly multitask, our ability to deeply understand and empathize suffers (Rosen et al., 2013). The brain begins to process signals in fragments, losing the ability to deeply connect. We communicate more often, but not necessarily more closely. We chase the form of communication and forget about meaning and content.

That is why it is important to remember: communication is not only a dialogue between people. It is a dialogue with yourself. With who we were. With who we want to be.

You will probably agree: even a simple lunch with a friend cannot be called a meal. It is an act of confirmation of connection. Sometimes we write a short "Where are you?", considering this phrase clichéd. But behind it lies a gesture of care, memory, and presence. Sometimes such words become the only reminder: "I remember that you exist."

Dale Carnegie emphasized: a sincere interest in another person is the basis of any strong connection. Communication does not begin with beautiful phrases, but with genuine attention, something that cannot be faked. A person always feels when he is seen and heard.

But attention is only the surface. Other mechanisms are at work underneath. According to Eric Berne, every communication we have has hidden scripts, invisible roles. We don't just communicate – we play, often without realizing the rules. In these games, **meaning is born not in words, but in relationships.** Even silence can be a signal that awaits understanding.

As research at the MIT Media Lab (Pentland, 2014) has shown, micro-signals—tone of speech, rhythm of movements, facial expressions—build trust faster than words. True communication begins with bodily attunement, long before the formal exchange of phrases. This proves that the field of interaction lives in us before meaningful speech appears.

But what happens when the connection with yourself and the world is broken? We can live in society, talk, work, be surrounded by people and still feel the absence of a real connection. And sometimes it's a complete loss. Internal disunity, when even loved ones do not hear, and words mean nothing. And then not just loneliness occurs. A failure occurs.

Human consciousness is designed so it cannot be satisfied with biology alone. It is not enough for us to perform functions. We want to understand where and why we are going. We need to understand, not just to exist. And when communication is disrupted, life begins to lose weight.

When thousands of people commit suicide every day, this is especially noticeable. It is not enough to be. You need a reason to be. If the inner question remains unanswered, everything collapses.

Viktor Frankl wrote, "A person can bear almost any how if he knows the why." (Frankl, 1946) This is why his logotherapy places meaning, not comfort, at the center.

And modern research confirms that a sense of meaning reduces anxiety, depression, and the risk of self-destruction (Heintzelman & King, 2014). **Connection and meaning are two sides of the same field.** Where one is, the other can be restored.

And this is where communication takes on its highest role: it doesn't just support life – it saves it. It gives it to us.

We don't just walk the road, we pave it with our questions.

Communicational Energy is not just a theoretical construct. It is an observable phenomenon supported by cognitive psychology, neuroscience, and social theory.

Research in psycholinguistics shows that when communication becomes deep, stable, and meaningful, it is accompanied not only by mental clarity but also by physiological shifts. For example, active connections reduce the level of cortisol, the stress hormone, and increase the level of oxytocin, the hormone of trust and attachment (Heinrichs et al., 2003).

Neuroscientists confirm that the human brain is literally wired for interaction. The mirror neuron system (discovered in the 1990s) is activated both by action and by observing others doing so. This is the basis for empathy, learning, and social coordination (Rizzolatti & Sinigaglia, 2008).

Even the perception of time changes depending on the connection. An important conversation can stretch the moment—creating a time-expanding effect (Block & Zakay, 1997). Communication affects not only what we are aware of, but also the structure of our perception of reality itself.

According to the information theory of personality (Danziger, Bruner), human self-awareness is not a fact, but a process. It is formed through internal dialogue. We become ourselves through stories about ourselves. And this story is built through communication with the outside world, with loved ones, with society, and above all with ourselves.

A lack of conscious connection with others has been linked to an increased risk of anxiety disorders, depression, and even premature mortality, according to research from the World Health Organization (Holt-Lunstad et al., 2015). Loneliness is perceived by the brain as a form of social pain, activating the same areas as physical pain.

Communicational Energy is the foundation on which each day is built. Every morning, when the screen of our reality turns on. Every word, every look, every gesture. And so, before returning to scientific models at the macro levels of physics, biology and cosmology, it is important to recognize: everything begins here, at the level of human life.

Communication makes us human. We exist because we are constantly connected. We are alive because we speak, listen, and feel. We move because we respond. And as long as this subtle but unbreakable field of connection remains between us and within us, we have a path. And the path does not begin with a goal. It begins with one simple act: being in contact. With yourself, with others, with the world. And that is everything.

COGNITIVE BLOCK - PRACTICE OF CONSCIOUS CONNECTION

1. Check the meaning of your words.
We speak thousands of words every day. But how many actually change something in ourselves or in another person?

Ask yourself:
"What have I said today that has helped me understand a situation better or helped someone else see something more clearly?"
If the answer is nothing, this is not a reason for reproach. It is a signal of direction: speak to change. Be silent to hear.

2. Listen with your body, not just your ears.
True communication is not expressed in words, but in the rhythm of breathing, in pauses, in intonations.

Try to observe for 20 minutes:
How does a person move? How does his breathing change? What does his rhythm say, besides speech?

Tuning the body is the first step to understanding without words.

3. Revision of connections: your social map.
Make a list of 5 people with whom you communicate most often.
Answer honestly next to them: "What is transformed in me when I talk to them?"

This map will show: where your energy grows — and where it is lost and blocked. Not all connections are equal. By choosing the environment, we choose the vector of our evolution. Where there is a real connection, growth is born.

Chapter 3 Summary

- Every action we take is a response to an external or internal impulse. We live in a constant system of cross signals.
- We do not exist in a void, but in a network. Even Monday, the route to the store or the morning gesture are formed by a collective setting.
- Search is a basic function of consciousness. We don't just react, we constantly ask – of others, of the world, of ourselves.
- Any question is already a form of connection. Even without an answer, it confirms: there is a connection, consciousness works.
- Connecting with yourself is not a luxury. It is a life support system. Without it, the meaning of life is lost.

Chapter 4

The Communicative Energy of Consciousness. What Makes the *I* Possible?

Until you make the unconscious
conscious, it will control your life
— Carl Jung

We use consciousness every day. But we rarely ask what it is. And yet it is what makes everything else possible.

What is consciousness? This is one of the oldest and still unresolved questions. On this path, humanity has gone from philosophical reflections to neuroscience and quantum hypotheses. But the answer remains shaky.

In this chapter, we understand **communicational energy not as an external phenomenon, but as an internal engine of awareness.** The brain works not only as a computing machine, but as a field of connections where millions of streams of signals, sensations, memories, and meanings are coordinated. Consciousness is not a fixed point, but the result of dynamic interaction: distributed processes, neural ensembles, oscillations, and switches.

Everything we call *I* arises in the process – not before it. We will consider how this internal communication works and why it can be considered a form of energy underlying our awareness. **We will try to prove that Consciousness is a mirror of the world and its architect.** It does not record, but distinguishes. It does not repeat, but rethinks. In consciousness, connections are born.

We are accustomed to thinking that Consciousness reflects what is happening. But it chooses what is happening in the first place. Thought, emotion, and choice are not bursts of reaction, but patterns of coherence into which perception, experience, and attention are woven. Consciousness does not live in the world. It forms the version of it that it can comprehend.

We already know that communication begins before language. But there is something more important here. Consciousness also begins not with thought, but with tuning. Human consciousness is not an isolated space. It is an open, dynamic system that continuously exchanges signals. We do not perceive information, we transform it into meaning. The brain does not function as a processor, but as a living, flexible, adaptive interface. It is where matter, energy, and information intersect. It does not limit itself to receiving and processing signals, it actively co-creates reality through the process of communication.

It often seems that thoughts are born inside us. But every thought is only the result of interaction: with oneself, with others, with reality. Even an internal monologue is a form of dialogue, turned inward. A person's strength is not only in intellect, but in the ability to connect meanings, find a common language, build bridges between different systems. That is why empathy, understanding, and sympathy are such powerful qualities. They are not so much social as energetic. These are forms of manifestation of Communicational Energy in its purest form.

We can transmit information and states. We influence each other with emotions, intonations, and speech tempo. This is confirmed by neurophysiological studies: when observing another person's expression of disgust, the same areas of the brain are activated as when experiencing it personally – as if the reaction were transmitted directly (Wicker et al., 2003).

Research shows that during a live dialogue, one person's brain literally tunes in to the other person's brain. Stephens et al. (2010) recorded that during communication, the same areas of the brain are activated in the interlocutors — this is called neural synchronization. Thoughts, pauses, intonations, gestures — the brain picks up on this and adjusts its processes to the rhythm of the other. That is why in joint singing, dancing, or ritual, there is a sense of we — a common rhythm, a common communication field. This is the work of Communicational Energy: transmission and tuning of several systems into a single one.

Now imagine you are in a dark room. You know there is something inside – objects, walls, maybe other people. But everything is blurry, you can't see the details. But then a window appears in the wall. Light comes through it. At first it is weak, then brighter. We see only contours, blurry outlines. Then brighter. The wider we open the window, the more light there is, the more clearly reality appears. Space takes shape. You see textures and colors. Light is consciousness. And Communicational Energy allowed the window to be built in and opened. It starts understanding. It brings the void to life.

Imagine a table in front of you. Three apples lie on it. They are motionless, and their position seems predetermined.

Now, a simple action. I pass you a note: "Move the middle apple to the right." You read, take the fruit and move it. That's it. The world has changed.

Neither force, nor instinct, nor biology caused this action. It was caused by the transmission of meaning. Hidden in the verbal code. But effective and realized. Here it is, Communicational Energy. We do not see it directly, but we see the result. Words transformed into intention, intention into movement, and movement into a new state of the system.

Communicational Energy is distinct from simple information transfer, performing the function of an active transformer of reality. The phenomenon of hidden Communicational Energy, or the energy of Consciousness, has significant transformational potential despite its invisibility.

Any person, with no knowledge or skill, can use this energy in a short time and get the desired result. Call a friend, read a book, find information on the Internet, and he can cook a soup that he has never cooked before. That is, until recently, he did not know how to do something, but, using Communicational Energy, he changed his reality.

But all this is possible only because the world responds to us with consistency. And here we first approach the key property of consciousness.

Consciousness cannot be reduced to a set of sensations or automatic reactions. It manifests itself as the ability to construct a holistic picture of perception. Being conscious means recognizing events, distinguishing objects and their meanings, and linking current experiences with memories. We see and become aware of what we see, so that in the future we can instantly recognize what is familiar.

This ability to recognize is possible because of the fundamental stability of the world. Green remains green, the moon retains its nature day after day. Even in endless movement, the essence of things remains recognizable.

If this were not so, we could not navigate reality. Colors and sounds would change without reason. Objects would lose their shape. Everything would disintegrate into a chaos of sensations, leaving no support for perception. Consciousness is possible because the world does not collapse in on itself every second.

Now try looking at a string of random characters:

апру емјпы один рыык пкв слы два соыоор begin

Even without understanding the meaning, your consciousness looks for the familiar. You notice: one, two, begin. There is no message in the chaotic line, but you are already trying to recognize the structure. This is how consciousness works: to connect, to distinguish, to complete. We literally peer into the chaos, hoping to find form in it.

Communication is impossible without a common code. And the more perfect it is, the more accurately one can convey the structure of the world. Therefore, language, symbols, and metaphors are not just tools of communication. They are ways of structuring reality itself.

Scientists suggest that if physical laws are universal, maybe the principle of coherence lies behind their stability. Consciousness is then not just our peculiarity. It is woven into the fabric of existence.

Albert Einstein believed that fundamental laws are preserved in any frame of reference. His idea of stability: even in motion and variability, there are constants.

He later lost a dispute with Niels Bohr, who claimed that reality does not exist until it is measured. It depends on the observer. But we will look at this dispute later in the book.

What is important now is something else: we can distinguish and organize perception. This ability is not given once and for all, it is developed. And it is possible only because the surrounding world allows for repetition and predictability. We learn to recognize: not just to see, but to recognize. This is the law of perception, the basis of awareness.

We like to think of consciousness as a flashlight illuminating reality. But it is more accurate to think of it as a spotlight painting the scene itself. You do not see as it is. You see how you connect. And what you do not connect does not exist for you. Consciousness is not vision, but selection.

It does not illuminate the world, but decides what will become the world. When we direct our attention, we do not notice – we form. Imagination in this process is not an ornament, but an active component. It projects possibility, and it is this that allows attention to choose and choice to emerge.

The formula for the manifestation of reality is simple: possibility x attention = reality.

Consciousness compresses infinity to the specific world in which you are now reading this line.

When ancient people looked at the stars, they didn't see physics. They saw stories. And it was these stories that transformed the chaotic scattering of dots into the constellations of the zodiac, predictions, and a map of the path. Humanity has never lived simply in reality. It has lived in a reality enhanced by imagination. And this is precisely the power of consciousness: not to reflect, but to interpret. Not to record, but to coordinate a picture of the possible.

Communicational Energy connects disparate points into lines, lines into figures, and figures into entire worlds. And in these worlds, meaning appears. Not chaos, not randomness, but a space in which one can distinguish, understand, and be.

Thus, emptiness becomes volume. Thus, movement turns into perception.

What we experience as *I* is a form in which the energy of connection has gained stability. Our *I* chooses one from billions of possible states and makes it reality. We do not simply live in the world, we assemble it in awareness.

Understanding is not a process of accumulation of facts, but a coincidence of the external with the internal, a point of agreement where a connection arises. This is where communication begins to work.

We may think we live in a stable world. But in reality, we can hold onto structure and give it meaning that creates this sense of stability.

There is a famous phrase: "If you want to change the world, change yourself." And it is not just philosophy. We must see in it the physics of perception. By changing the settings of attention and interpretation, we build the very model of reality in which we find ourselves.

Consciousness did not suddenly appear. It was not given to us by the Gods. It evolved — as a survival strategy, as an adaptive system for navigating an increasingly complex reality. Primary forms of self-awareness, according to biologist and cognitive scientist Michael Gallup (Chimpanzees: Self-recognition. 1970), can be found in chimpanzees, dolphins, elephants — they recognize themselves in the mirror. This is not just image recognition — it is a signal: "I am I."

We do not see the world. We distinguish what is agreed upon within us.

Neurophysiology shows that as the neocortex, especially the prefrontal cortex in mammals, develops, the functions of planning, empathy, and reaction delay become stronger. These qualities are directly related to what we call mindfulness.

According to the social brain theory (Dunbar, 1993), the mind evolved primarily as a tool for maintaining stable relationships. To understand others, we first had to learn to understand ourselves.

Consciousness cannot be called a passive mirror. It functions simultaneously as a projection screen and an experimental laboratory: it creates internal models of reality and continuously tests them for viability.

Global workspace theory (Baars, 1988) speaks of consciousness as an integrative structure: memory, perception, attention, all come together at a single point of assembly, in a single cognitive process.

Experiments by Roger Sperry and Michael Gazzaniga (1960–70s) confirm this principle. In studies with patients suffering from severe epilepsy, scientists surgically cut the corpus callosum, the neural bridge connecting the hemispheres of the brain. The result was astonishing: the brain began to function as two autonomous systems. Paradoxical situations were seen: the left hand acted independently of the right, the patient could simultaneously button a shirt with one hand and unbutton it with the other.

These data convincingly proved that consciousness is not a monolithic formation, but arises because of integration processes. It is enough to separate the communication channels, in this case neural connections, and the single *I* is divided into two independent streams of awareness.

Such scientific examples lead to conclusions: **consciousness does not arise from the center, but from integration. From interaction and communication.**

And if internal communication creates coherence between parts of the brain, maybe it also creates coherence between consciousnesses in external, interpersonal communication.

Then human communication turns out to be an exchange of signals and a continuation of the principle of consciousness itself. In dialogue, the boundaries of the *I* expand, creating a common semantic field that influences the awareness of reality by each participant in the interaction.

Modern theories confirm this. In integrated information theory (Tononi, 2004), consciousness is defined as connectivity of information in a system. The higher the integration, the higher the level of awareness. In this model, consciousness is not yes/no, but a spectrum: from primitive forms to complex self-awareness.

In the predictive theory of the brain (Friston et al., 2009), perception is not seen as a response to an external signal, but as a forecast that the brain constantly checks against incoming information. Consciousness is the coordination of expectation and experience. It works as a continuous attunement mechanism, and this is where Communicational Energy manifests itself.

Even at rest, the brain forms stable patterns of activity (Koenig et al., 2022). These are neural patterns that ensure continuity of perception and internal connectivity. Max Tegmark (Life 3.0, 2017) calls the brain an "ocean of active information." It stores data and patterns of interaction—communication structures that define who we perceive ourselves to be.

Consciousness is a flexible dynamic system without rigid boundaries. Modern research (Pascual-Leone et al., 2005) shows the ability of new experiences and knowledge to change the neural architecture of the brain. The process of perception is constantly transformed under the influence of experience. At the same time, communication gives these changes meaning and direction.

Consciousness develops according to the principle of a living organism, where information flows perform the function of building blocks, and communication processes set the structure of this continuous renewal. Just as different buildings are built from identical bricks, and different patterns on roads were laid from identical cobblestones, so we lay and build roads and the architecture of our lives.

Thus, Consciousness became the turning point of evolution. The tool that allowed life not just to survive, but to understand. Not just to react, but to predict. Not just to change, but to model.

Consciousness is not a fact, but a function of complexity. It holds and coordinates multiple levels of interaction. In this sense, **Consciousness = precision and depth of the coordination tasks being solved.**

Humans first put consciousness above evolution. For millions of years, nature has been honing forms through natural selection. But now we don't wait for the environment to select the best.

Through selection, technology, and genome editing, humans have begun to design life themselves.

This has borne fruit, we increase crop yields, control reproduction, and prevent diseases. But along with this, strange consequences have appeared. There are breeds of cats with huge eyes and flat muzzles, and their faces seem cute. But they have a serious breathing problem because of their short nose. There are tiny decorative dogs that cannot give birth without a cesarean section. We created an idea and subjugated the body to it.

Now form is subordinated not to the body, but to the idea. Not to the environment, but to the inner image. Consciousness does not follow evolution. It directs it.

Consciousness is the point where the possible takes shape

Physicists say: at the quantum level, the world is a potential, not a fact. Particles do not have a definite state until they are recorded. The world exists as a possibility until the moment of an act of attention, an act of connection. It becomes reality when an observer appears. When awareness arises.

This principle works not only in the microworld. And not only in the laboratory. It works in us. Consciousness is not just a way of perception. It is a system of structuring existence.

In neuroscience, this is described through *bottom-up models* — ascending processes: from a sensory signal to a pattern, from a pattern to a conscious structure. Perception is formed from the bottom up: from simple sensations to complex meanings. In artificial intelligence, we see similar logic: first, raw data, then the identification of patterns, and then the formation of meanings and behavior patterns.

My cat Mozart asks to play the same game every night. I hide my hand under the blanket and move it. The cat switches on the hunter mode and rushes at the ledge in the blanket, as if it were prey. But the main thing is that he intuitively understands: the real thing is not on the surface, it is under the blanket. And he tries to get to the hand to what is hidden, but causes movement.

So it is with us. Sometimes we look at the world and perceive only the form of matter. But underneath it is the movement of Consciousness. We do not always distinguish what is moving beneath the surface.

Matter produces motion. But matter is only a shell. The outer side of the carrier. When someone runs in a tracksuit, we don't say that the suit is doing sports. Or that the tailcoat is going to the theater. The carrier does all of this.

The shape of the body, the features of the face, the structure of the ears and nose, even success in life, all this is not the merit of a piece of matter. It is a manifestation of consciousness inside the body, inside the DNA, in the structure of the environment, family, culture, coordinate system. Consciousness has adapted and developed a form ideal for its existence.

Have you ever thought that even the architecture around us is a trace of communication?

Imagine: the Cathedral of Santa Maria del Fiore in Florence. Or Notre Dame in Paris. We think that they are like that on their own. But these architectural masterpieces arose from the connection between the customer, the architect, time, engineering thought and materials. If something in this system were different, the building would be different. Many cathedrals were rebuilt dozens of times. Religion changed, and the form followed it. A new meaning appeared, and the material shell was rebuilt. Form is not an absolute. It is frozen communication.

Consciousness has three key characteristics:

- filling,

- interaction,
- constancy.

It gives meaning, seeks connection with other consciousnesses, and stabilizes the laws within the world. If these principles did not work, the world would fall apart. There would be no logic, no differences, no structure. But consciousness makes the world manageable, connected, and meaningful.

If these functions of consciousness disappeared, the structure of reality would disappear. Consciousness does not observe, it brings order. It does not record, but forms. It is what turns probabilistic noise into a meaningful picture.

Consciousness is not static. It moves, explores, and strives for the boundaries of the possible. It encourages us to discover new things, overcome obstacles, and create.
We see an abyss and start building a bridge. Having seen the sea, we invent a ship. Consciousness is always in search of forms, meanings, and directions.

Maybe this aspiration holds the key to understanding reality. It can be assumed that reality is formed at the intersection of two types of fields: the first, which has the potential for directed interaction, and the second, which can take certain forms.

The basis of the world is not only the manifested, but also that which is in a state of potential manifestation. Communicational Energy is a directed field of possibilities. Dark Matter serves as a medium for the emergence of forms. Consciousness, acting in material reality, becomes the place of their meeting, the border where the formless takes shape. It is here, at this point of intersection, that what we call "the world" is born.

We will talk in more detail about Dark Matter and its connection with our hypothesis in the next chapter.

In this chapter, we tried to understand Consciousness as a driving force capable of organizing and structuring reality, connecting matter, energy and information into a single whole. As a form of manifestation of Communicational Energy.

A form in which matter begins to realize itself.

And man is the voice of the Universe, pronouncing not only "I am," but also "I am aware of myself."

🧠 COGNITIVE BLOCK - CONSCIOUSNESS AS A WINDOW INTO REALITY

1. Trace a thought to its source

Stop at one thought that came to your mind today. Any about yourself, about others, about the world.

🔹 Ask yourself: "Is this mine? Or is this a reaction to someone? Is this thought a product of experience or influence?" Consciousness is a filter. But you can observe what it lets in.

2. Feel the moment of neural synchronization

During a conversation, choose when you are fully engaged, listening with all your attention.

🔹 Record it physically: "Am I relaxed? Am I tense? Am I mirroring the gestures of the interlocutor?" This is the synchronization of consciousness.

3. Practice the act of creating meaning

Take a random object (item, photo, word). Look at it for 30 seconds.

🔹 Ask yourself: "What am I seeing now, form? Or meaning? "Now change the context, add a story, an emotion, a name. Look again. Consciousness does not wait for meaning. It creates it.

Consciousness is a light that chooses what to illuminate. But you can learn to control its direction. And then the world becomes not only understandable, it becomes yours.

Chapter 4 Summary:

- Consciousness is an active system that shapes perception and reality.
- It arose as an evolutionary mechanism for survival, modeling and prediction.
- Neuroscience confirms that consciousness results from complex integration of information, prediction, and synchronization.
- Communicational Energy through consciousness not only transmits signals, but transforms and structures external and internal models of the world.
- Empathy, synchronization, and neuroplasticity are the biological foundations that make communication deep.
- We are conductors of Communicational Energy, creating order, meaning and form.

Conscious point of reality = attention * field of possibilities

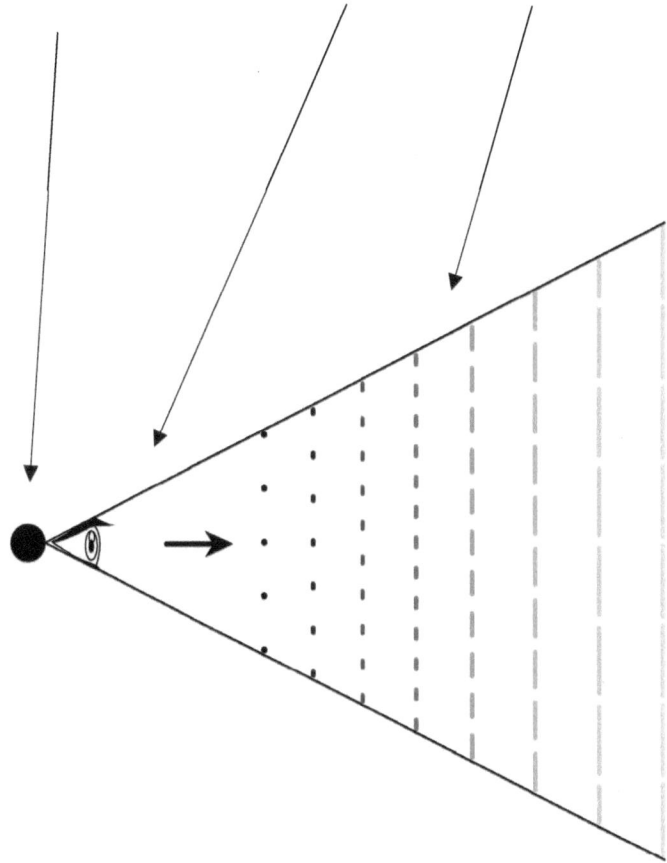

Vision of reality

CHAPTER 5

TIME AND THE HIDDEN FIELD OF POTENTIAL MATTER

*I know not what I appear to the world,
but to myself I have always been but a
boy, playing on the shore of a vast
ocean of truth — Isaac Newton*

When was the last time you were at the theatre? Which row did you buy tickets for? One of my friends joked: "Some people buy tickets in the front row not because they are closer to the stage, but so that all the other spectators will notice them there."

In ancient Greek theaters, everything was thought out so even the last row heard and saw perfectly. Without microphones, without amplifiers, 2000 years before the invention of speakers. There, in the upper tiers, the most noble sat, to be *above everyone* else and to survey *everyone*.

But the place in the hall is not so important. The main thing is that you *see and understand* what is happening. Theaters have always been judged not by the seats. The quality of the stage and the actors' performance have always been admired. Only this could create a great theatrical performance.

Have you ever wondered where the need for theatre came from? It may be originally embedded in people to recreate patterns of life, to reflect the movements of the world and to observe them with delight. (But about beauty and its deep nature in one of these chapters.)

For now, let us remember the great phrase of the great playwright William Shakespeare:

"All the world's a stage, and all the men and women merely players."

In earlier chapters, we have looked at the main actor in the Universal performance – Communicational Energy. We defined our consciousness as an attentive spectator, and most likely as a director of the performance, observing the action from the auditorium.

Now it's time to consider the **stage**.

Any stage is famous for its volume and the ability to change scenery. The greater these indicators, the more grandiose the show can be.

Now imagine that you are sitting in a theater, and the stage in front of you is limitless. Although it is difficult to imagine, let's simplify: let it go a hundred kilometers deep and stretch several kilometers wide. A grand scale, capable of accommodating everything you can imagine. There is enough space for any scenes and decorations.

But this is not enough. This space should be alive, so that the scenery does not freeze, does not interfere with the next act, that the action continues. Even the most magnificent spectacle turns into a burden if it loses its mobility.

And here the key mechanism of change appears on stage: entropy. The organizing principle that allows the performance to move forward.

Entropy is not just a physical term. It is a universal law of the disappearance of differences, a measure of disorder. Everything in the Universe, from dust particles to galaxies, is subject to this law. The movement from order to chaos, from form to decay, from structure to dispersion. What yesterday seemed solid and stable, today is already changing, and tomorrow crumbles and disappears.

In physics terms, entropy is a measure of probability. The system loses distinctions, unique forms dissolve, and ordered states give way to more probable but less distinguishable configurations. Everything strives for maximum probability and minimum information content. As entropy increases, everything becomes less and less certain. Distinctions are erased. Forms disappear. Signals fade. Information fades into uniform homogeneity.

So maybe the final stage of entropy is chaos and a return to the original background? To a state in which there are no differences. No meaning. No form. There is only potential and no manifestation.

When I was a kid, I had a toy called *Magic Screen*. Two rotating wheels controlled a thin line: one moved it up and down, the other left and right. You could draw a house, a ship, a labyrinth on the screen, anything you wanted. But the most magical thing happened later: you had to shake the screen, and it became clean again. Everything disappeared. And you could start over.

This is entropy. Not destruction for destruction, but the **zeroing of form to free the field for a new one.** Disorder that gives a chance to a new order.

Now stop for a second. Look at your life as a change of scenes. What scenery have you already changed? What do you regret that has dissolved into the past, and what are you grateful for that has gone?

Maybe entropy is not our enemy. It is our invisible ally in the art of renewal. Through it, we not only lose, but also gain. It is entropy that makes change tangible – **and thanks to this we feel the flow of life.** We distinguish before and after only because the order changes. And with it comes the sense of time.

This everyday experience of feeling everything change reflects one of the most fundamental laws of nature. Modern science also sees entropy not just as chaos, but as a direction of development. Roger Penrose described the arrow of time as a psychological, thermodynamic, and cosmological direction, all combined into one inexorable vector: we cannot go back. Stephen Hawking wrote that at the moment of the Big Bang, entropy was minimal, and since then the world has only been falling apart. *(Penrose, "The Emperor's New Mind," 1989; Hawking, "A Brief History of Time," 1988)*

If it is scientifically proven that everything that is born in the Universe usually disintegrates and that time, the fourth dimension of our reality, moves in one direction: from order to disorder, then isn't this direct evidence that our reality interacts with some fundamental force whose nature is destruction, disintegration, defocusing, disordering?

In our concept, this can be called **dark Nothing**. Not darkness. Not evil. But a field where everything could be, but has become nothing yet. Or was, but has become nothing. Like a blank sheet of space in which events can occur, but which itself does not take part in any of them.

It can be assumed that the potential field we have described is the phenomenon that science calls Dark Matter. We cannot detect it directly: it does not emit light, does not interact with the electromagnetic field. But we observe its influence, by the gravitational distortion of light, by the way it keeps galaxies from disintegrating. It is thanks to the effect of gravitational lensing that the Hubble and Webb telescopes (NASA, Hubble - "Dark Matter and the Cosmic Web", 2021) build maps of the invisible: where light is refracted without a visible source of mass, we record hidden matter.

According to these observations and scientific calculations, Dark Matter makes up about 27% of the entire mass-energy of the Universe. Visible matter, for example, is only 5%. The rest is occupied by Dark Energy, something even less understood by scientists. But it is this, according to modern theories, that is associated with the accelerated expansion of space.

Dark Matter is not an object, but a **condition**. A silent but structuring phenomenon. We do not hear it, but perceive it as pauses between sounds by their influence.

Modern physics has long since changed its view of emptiness. **Vacuum is not silence, but a quantum storm,** where virtual particles continuously flare up and disappear. This has been confirmed by both experiments (Casimir effect, Lamb shift) and quantum calculations.

The so-called nothing is a hyperactive fabric of possibilities, pulsating on the edge of being. This explains the position of modern scientists: the field is primary, while particles are only excited states of this field.

Matter is the possibility of connection that has become form.

We pronounce the word reality and almost automatically divide it into three components: past, present and future. We believe in this axis of time as something indisputable: we live here and now, we move forward, and we consider the past to be finally completed. But how does the present become the future? Why is one moment fixed, while another remains only a possibility?

This forces us to rethink the nature of time – maybe it is not an objective coordinate, but a subjective effect of the consciousness's perception of changing material states.

So, in putting forward our hypothesis, we assume that the nature of the future can be understood if we look at it through the prism of the Dark Matter concept we have proposed. Not as an empty expectation, but as a probabilistic potential field. The future is not something that will automatically come, but something that can take shape in a specific implementation form. Where a set of decisions and choices forms the trajectories of the possible. We do not see it directly, like Dark Matter, but we feel a connection with it through our own actions and choices, which form implementation trajectories and give direction to the perception of time.

This model opens up prospects for further study of the relationship between consciousness, time and fundamental physical processes.

Each intention, each choice is like an impulse sent into the potential. The future has no form yet, but possibilities are already encoded in it. It is as if we are moving through an invisible fabric, where Communicational Energy, our directed force, activates hidden possibilities, making them reality.

Every thought, every decision seems to be a tunnel into the future. Like a 3D printer forming a new form layer by layer. The closer the moment of implementation, the denser its contours become, the more tangible and distinguishable the future reality becomes.

We often plan our summer vacation in winter. We cannot yet see ourselves lying on the beach. But by buying plane tickets and booking a hotel, we have laid the first threads in the future reality. If the plane leaves on time and is not cancelled, this reality will manifest itself. Six months ago, it was only a possibility. And now it is becoming a body, an event, a fact.

We live on the border of two worlds: memory (realized matter of the past) and potential (not yet manifested matter of the future). And between them, a transition is activated not by time, but by communication. And the Communicational Energy activates the transition between them.

What does exist mean? What do we call reality? Usually, something that can be seen, touched, or measured. But maybe reality does not exist in itself, but something that is made possible when two invisible forces are aligned.

Reality arises because of the coordination of two invisible potentials.

On the one hand, there is the Communicational Energy. This is the logical field of action: it directs, differentiates, and offers structure.

It is the ability to connect everything with everything, but without manifestation (since it cannot manifest itself – we can call it Dark, Hidden Energy)

But there is Dark Matter. Dark because it is not yet manifested, not yet visible. It is a potential medium of embodiment: formless, but responsive. It stores everything that can be, but without its own impulse.

Neither creates the world separately. Communicational Energy sets the direction. Dark matter offers capacity. But when the directed impulse enters into readiness to be form at the point of contact, what we call manifested being occurs. A wave appears, an oscillation that we perceive as form. Everything that we perceive as things or events are patterns on the surface of this contact.

We live in two layers of reality at once. You can imagine it like this: a person stands in the water with a life-saving vest on. His body feels the water, his feet feel the bottom. He moves, feels the current, the direction. But the head is above the surface. It sees everything that is under the surface of the water.

Consciousness is in a special position. It is connected to the body, but no longer in water, not within the environment.

It belongs to another layer. So we simultaneously exist in two worlds, in the visible and the invisible, in the manifested and the potential.

The material layer is action, density, and form. The potential layer is the field in which forms become possible. And consciousness is the point of connection between them. It acts as a communication node.

Matter does not arise by itself, but because of interaction in this field. And consciousness is not an external observer, but a trace of interactions that have occurred. It reflects the connection.

In material reality, we shake hands. And in communication, there is an impulse, intention, attention that trigger this movement. First, a signal arises. It passes through the field, turns into action, is fixed in consciousness and generates the next impulse. This is how a circle arises: .. → field → matter → consciousness → field… Each element influences the others. Nothing exists separately. This is not a philosophy, but a description of the behavior of systems in which each act of communication changes the entire structure. What we feel, do, notice returns to the field and changes its state.

This idea follows the quantum interpretation, in which observation affects the state of the system (Tegmark, 2014; Rovelli, 2021). We add one step: consciousness does record reality and becomes part of the feedback. It reconfigures the field, giving impulses direction, meaning, and stability.

Everything that happens to us and around us is a trace of the configuration of the connection between Dark Matter and Dark Energy, which has achieved stability and formed into an event.

In this sense, reality is neither a stage nor an object. It is a continuous process of coordination. If we imagine that behind all these processes there is an internally ordered field: communicative, dynamic, algorithmic, it becomes clear: form is not born. It results from a change.

When we use the term algorithmic dynamic communication field, we are describing its functional properties.

Algorithmic - because this field is recalculable. It carries an internal logic that allows for the configuration of connections to be updated with each new interaction. It is a computable, adaptive space, not a fixed structure.

Dynamic - because it is constantly changing. Any impulse, any connection causes a transformation in the state of the field. It is not static and does not retain one form, but is constantly being rebuilt.

Communicative - because all elements in this field are connected to each other. The connections between them are not external, but built into the structure of the field itself. And these connections determine how the structure, movement and form of what we perceive as visible matter will change.

It is not a background for interactions, but an active system in which each act of communication changes the overall configuration. Field algorithms are not fixed, but adaptive. Each impulse launches an algorithmic wave, recoding the entire pattern of reality. The field feels itself through interaction, records the response, and is recalculated. We do not carry the essence within us, like a static load. We are a configuration of communication, instantly updated by the system. Each interaction is an event and a transformation of the entire structure of the field.

In the workshop of the Universe, the canvas is woven from Dark Matter. And the brush is formed by Communicational Energy. Consciousness is just a point where their touch creates a picture of the world.

The axiom of choice in mathematics states that one can choose an element from any set, even if the set is infinite and has no explicit rule for choosing. This idea seemed abstract for a long time until one of the most paradoxical, but mathematically correct, theorems of the 20th century appeared: the Banach-Tarski paradox.

According to this theorem, a sphere can be decomposed into several non-intersecting parts, and then, by changing their position in space, two spheres of the same size as the original can be put together. Without adding matter. Without violating mathematical laws. For many scientists, this was a challenge to common sense, but from the point of view of our hypothesis, this is only confirmation: reality can be reassembled if the interaction with its algorithm is built correctly.

It would be nice if such laws worked in everyday life: take a wad of money, arrange the bills specially, and collect twice as much. But in theory, we get infinity from one volume, but in practice, everything disappears somewhere from one wallet.

Consciousness in communication acts as a node connected to infinity. And that is why it can choose. Just as in the axiom of choice, we can extract an element from an infinite set without an algorithm, so consciousness extracts a form from the field of possibilities, even if it is not yet defined.

This field is not linear. It does not follow a time axis. Rather, it can be thought of as a **closed energy structure**, where any point can be a beginning, an end, or a transition.

We don't know yet exactly what form this field has. But the closest image could be a **torus**, a structure where energy returns to itself, but at a new level, having undergone a transformation.

In Chapters Nine through Eleven, we will look at models of algorithmic interaction and possible scenarios for transition between them.

Consciousness, interacting with the potential structure, can project one of the states into reality. Without violating its integrity, it reconfigures its configuration.

We don't just observe. We influence, we adjust, we start. One communication point impulse, and the entire field can be filled with new meaning, like a reassembled ball in the Banach-Tarski paradox. Everything is the same, but different.

This is the algorithmic essence of the communication field: it does not dictate, but responds. It is not static, but can be formed. Reality is not set once. It is a response to attentively including consciousness in the fabric of the possible.

With this understanding, we do not observe the world, but are already included and manifested within it. And what we call ourselves is not an autonomous object, but a node of communication, a stable combination of field lines. And reality is not a frozen result, but a current configuration of coordinated interactions.

Our concept has direct correspondences in modern physics in how the states of quantum systems are described.

The potential field in quantum mechanics has its analogue under the term Hilbert space. Not space in the usual sense: it has no coordinates, but it has an internal structure. Directions, distances, angles are possible in it but not positions. Each state of a quantum system is a vector in this space. The system itself exists as a superposition, the sum of all possible states. But it manifests itself only at the moment of connection: when interaction, measurement, and focus of attention occurs. Then a multitude of options are projected into one state.

Manifestation does not arise from emptiness, but from structured potentiality. Connection does not create form from scratch; it activates what is already possible. In this understanding, form does not exist as an object, but as a stable disturbance on the border between being and potential. Matter expresses the coordinated state of the field. Consciousness manifests itself as an algorithm of coordination, born in the interaction itself.

Modern science is bringing us ever closer to understanding: observation changes reality itself. In quantum physics, this has been confirmed experimentally. Until a measurement occurs, a particle exists in a state of superposition, many possibilities. But the act of observation chooses one of them, forcing the potential to become actual.

And here we are not talking about the devices in the experiment, but about the fact of interaction: **reality manifests itself only at the points of connection**.

We have also put forward a hypothesis of a communication algorithmic dynamic field as the basis of our reality. But the following question arises: at what speed do such transformations occur? We perceive them as a continuous flow, as real time. And what parameter can describe this speed in scientific terms?

There is one universal limitation in the theory of relativity: the speed of light is the maximum speed of propagation of any signal or interaction. Neither matter, nor information, nor even the impact of an observer can pass through space faster than this value.

We assume this is not just a physical limit. It is a structural limit of reality itself. The speed of light is the speed of transformation of reality under the influence of the communication field. The maximum rate at which the potential becomes actual, and reality can unfold as a response to the impulse of communication.

Who decided that reality should be measured by the speed of light? Why not by the speed of sight? Or call it the speed of observation? At what speed do we let reality into ourselves? Or more precisely, at what speed do we create it as observers?

Light is an abstract stream of photons; no one knows where it came from or where it was hanging around. And a look is a concrete act, with tangible consequences: you looked, and reality appeared. You didn't look, and according to quantum physics experiments, it was as if it didn't exist.

Scientifically, light is objective. But from the point of view of your consciousness, your eye is more objective. It is closer. It is yours. It creates what you call "there it is."

The speed of light is a limit and the maximum frequency with which the Universe renews itself. It is as if each point in the field is given permission to manifest itself, but no more often than once every 299,792,458 meters per second.

In our hypothesis, the world unfolds in quanta of coherence. This makes time not a river, but a series of flashes. And if one of them is your attention, then it already has an act of creation. And in this understanding, it is not about how fast a photon and light as an object move. But about how quickly a form can manifest itself if attention, a signal, an interaction is directed at it.

Modern physics provides an interesting confirmation of our hypothesis.

The Italian scientist Donatello De Serafino developed the theory of elementary cycles (Elementary Cycles Theory, 2023), according to which each elementary particle is not a point in space, but a periodic process living in its own rhythm.

In this model, time is not a continuous flow, but a network of ultra-fast repetitions, within which all physical processes unfold. Reality does not manifest itself, but at the moment, the internal rhythm of the observer coincides with the rhythm of the event.

This is remarkably close to our hypothesis: the world emerges through the alignment of attention and possibility. What you call now is a point on a timeline and a flash of coincidence between you and the system, between intention and form.

If you still think this sounds too bold, here's what science says:

- In the theory of relativity, the light cone limits the area of possible interactions, everything outside of it is fundamentally unattainable.
- In quantum physics, reality is recorded only at the moment of measurement, and this recording cannot happen instantly, it is subject to the limitation c (c is the maximum speed of signal propagation, equal to the speed of light in a vacuum).
- In information theory, no signal can convey meaning faster than c, which is the basis of the law of causality.
- In the philosophy of consciousness (von Neumann, Wigner, Penrose): the act of observation is not a passive registration, but an active transformation. Consciousness does not simply reflect the world, but collects it from many potentials, and does so at the same speed of unfolding.

The communication field acts as a structuring agent, but its manifestation in time is limited by a physical limit: the speed with which the possible can be caught by consciousness and fixed as a fact.

We don't just observe the world. We activate it. But even activation is subject to a universal limit. Even unimportant elements such as attention, perception, intention have the same limitations as matter.

The projective nature of reality resonates within the framework of advanced scientific theories. For example, we can find a similar idea in Juan Maldacena's (1997) hypothesis of holographic correspondence (AdS/CFT), which changed modern theoretical physics. It states that reality in a space with gravity can be described by a theory on the boundary of this space, without gravity. This means that everything that we perceive as volumetric, dense, physical reality can be a manifestation of more fundamental connections and deep logic at another level. Form and substance become not the basis, but the projection of a coordinated information structure. Connection is primary. Geometry and matter are consequences.

But these principles are not only manifested in gravity models. At the level of quantum physics, they are confirmed experimentally in the behavior of elementary particles. An electron does not have a fixed position until it is measured. It exists as a cloud of probabilities. It appears in a specific position only at the moment of observation or interaction. Until we look, it is a possibility. This is not an assumption. This is physics.

In the scientific world, an important piece of evidence that reality exists only at the moment of observation was the famous EPR paradox (Einstein-Podolsky-Rosen, 1935). Einstein tried to show that quantum mechanics was incomplete, arguing, "Reality must exist whether we look at it or not."
But experiments, beginning with the work of Alain Aspect (1981–82), have shown that the state of a quantum system is fixed only at the moment of measurement. Until there is an observation, the world remains in a superposition of possibilities.

This effect (violation of Bell's inequalities) has been confirmed many times, and today it underlies quantum technologies. → Reality does not exist by itself. It manifests itself through the act of communication between the observer and the potential field. Isn't this evidence that matter itself is not a fixed body, but a function of communication with a potential field.

If we rise to the level of interpretations, we can see that the dynamics of reality selection are already described in physical models, especially in Richard Feynman's approach (*Path Integral Formulation*), which underlies quantum field theory.

He proposed an unusual idea: a particle — for example, a photon — does not choose a route in advance in its movement from point A to point B. The theory suggests that the particle explores all possible trajectories simultaneously. And each of them contributes to the final result with a certain weight. In reality, not any of them is realized, but the one where the total contribution is most consistent.

The space around us turns out to be not an arena for a single course of events, but a field of all probable trajectories. Reality emerges only at the moment of interaction, as a choice of one most balanced option from the entire set of possible ones.

This concept is close to our idea of a communication field: a space of potential matter in which the connection between the possible and the manifest is realized through the instantaneous coordination of multiple scenarios.

The reality we know is not the only possible option. It emerges because of selection, a collective superposition folded into the final frame of existence. The most coherent choice from countless probabilities.

We do not live in time, we move the possibility of becoming reality.

Think about how often you have said the phrase "I don't have time." We don't always notice that behind this trivial phrase, there is more than just a lack of hours. It means: the volume of our reality is already filled. We cannot accommodate something else, a task, a meeting, an experience.

We often complain that life flies too fast. And we try to systematize the chaos of time somehow. We came up with calendars, schedules, reminders, everything to somehow capture and record time. But it still slips away. Maybe because time itself is not what it seems?

Modern physics does not describe time as a separate entity. In the equations of fundamental theories, it is absent or appears as a derivative effect. Einstein showed that time is not an absolute, but a flexible coordinate dependent on speed and gravity. In quantum mechanics, it becomes blurred, and in cosmology, it acquires direction only because of entropy.

Physicists like Carlo Rovelli believe that time is not fundamental. In his book "*The Order of Time*," he argues that at the most basic level of reality, time does not exist—it only arises where there is consciousness and the ability to distinguish before and after. At the most basic level of the universe, everything exists. No clocks. No chronology.

The paradox of time is exacerbated when we look at the equations of quantum physics. They are all reversible. There is no preference: forward or backward. The arrow of time is not a law, but a consequence. Some physicists conduct experiments in which a process can be locally reversed in time, provided it does not interact with the outside world. This means that the world itself does not age; it ages only when observed.

The act of communication, the act of fixation, is the birth of time. And again, everything comes down to interaction, to transmission, to the energy of communication.

And this changes the focus: we perceive time as a flow, but maybe it is not time that moves. We move through a motionless present that exists forever, like a frame of a film where only the one watching it changes.

Over half a century ago, the phenomenon of superfluidity was discovered. At extremely low temperatures (below 2.17 Kelvin, approximately -270.98°C), helium loses internal friction, at which point particles begin to move as a single quantum whole. This liquid can flow down the walls of a vessel in any direction, and through microscopic cracks without resistance. This effect is not just a curiosity. It suggests that at the level of matter, a form of absolute coherence may exist, where friction disappears and movement becomes free.

Helium is one of the most common elements in the Universe. It represents matter with energy, visible, real, alive. But if helium in a superfluid state loses the individuality of atoms and becomes something like a quantum fabric, then why not suppose that Dark Matter is a form of even denser and more homogeneous substance? Absolutely cold, without friction, without information, without differences. Frozen homogeneity. Absolute potentiality. Without signs of anything except itself.

Now, imagine a telescope. When folded, it is short. When unfolded, it is long. But it has no fixed length. There are only proportions and the focal distance between the lenses.

The same is with the age of the Universe. It can be 13.8 billion years, or one moment. It all depends on the focus. Whoever observes determines the scale.

We do not live in time; we live in a process. Everything that happens to us is not a movement in time, but an opening up of possibilities. Transitions of states.

A person does not simply exist in time. He literally feeds on it. Any action, any thought, any experience is processing. We take a portion of time and turn it into an event. Into an experience. Into a meaning. Transforming potential nothing into something real.

Life goes on as long as there is Communicational Energy within us – to move, to differentiate, to connect. We live as long as something in us can split time into actions, thoughts, and interactions. When the energy runs out, we freeze and stop interacting in the potential field. We freeze to eternity. We dissolve in motionless potentiality. We remain... in Time. But no longer in the flow.

We are not just particles of matter. We are active lines in a field through which the energy of connection passes. This energy organizes, records, and transmits. We leave traces behind us: in DNA, in language, in culture. We carry information, moving through time and creating form.

In the next chapter, we will trace how this energy, which we have defined as Communication, participated in evolution: from molecules to cells, from instinct to reason. We will consider how the ability to coordinate developed, and how the structure of thought, memory and consciousness gradually grew out of the exchange of signals.

Chapter 5 Summary:

- Time is not an absolute, but a perception that arises in the interaction of consciousness with change. It does not move, but differs.
- Dark matter acts as a stage, as a field of potential and probability, and not as a thing.
- We do not live in time, but in manifestation, reality unfolds through acts of connection.
- We do not live in time, but in manifestation, reality unfolds through acts of connection.
- Consciousness does not record what is ready, but adjusts the field configuration and launches a wave of manifestation.
- The speed of light is a parameter of physics and the maximum rate at which potential becomes reality.

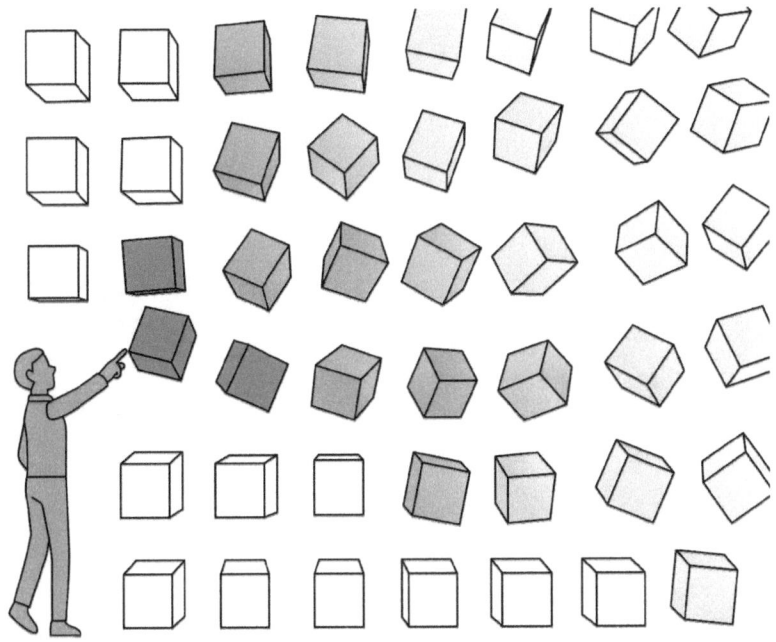

Single action and algoritmic dynamic communicational field

CHAPTER 6

MAN ON THE PATH OF EVOLUTION

Evolution is the unfolding of
consciousness to its fullness — Ken Wilber

In previous chapters, we have traced the hidden logic of the Universe - how connection shapes matter, how interaction creates consciousness, how Communicational Energy becomes the field in which reality takes shape. We looked into physics, cosmology, and neuroscience to see that everything living and nonliving exists thanks to the ability to act in concert.

Now it's time to take the next step: to shift our focus from macro levels to biological evolution. How exactly does the connection built into the structure of matter begin to form complex living systems? How does communication become not only a mechanism of transmission, but also a means of survival?

Any communication is an interaction, whether within an organism, between organisms, or on an ecosystem scale. At the core of any such interaction is the desire for stability, coordination and development. Consciousness does not arise from emptiness. It grows out of a dense system of connections. And the connections themselves grow out of exchange, reaction and tuning.

In this chapter, we will try to trace how communication developed in the process of evolution, from a molecular signal to collective behavior, how consciousness became an instrument not only of reflection but also of coordination, and how communication turned into a biological code of survival.

Evolution is not only the selection of the strongest, but also the success of those who discriminate more accurately, transmit more effectively, and unite more deeply. Life on Earth is not a random experiment, but a natural complication of forms, mechanisms of coordination and transmission of information. From the first molecules to quantum computers, everything follows a single vector: the vector of communication.

How did it all start?

Imagine: you look into a pot where something ancient is boiling, bubbling, seething - but no, it's not soup. This is what scientists call the **primordial soup**. A mix of molecules, water, temperature and random thunderstorms. Not very appetizing, but historically significant.

And it is in this hot, saturated environment that the **first amino acids appear**. Formally, they were just chemical compounds. But in essence, they were **carriers of potential for communication**. They weren't just hanging in space like spices in a broth. They began to connect. To react. To integrate into each other. To build chains.

This was the **ground level of communication**. Without words, without thoughts, but already directed. It was the ability to connect in a certain sequence that allowed molecules to form proteins, and those, in turn, laid the foundation for the first cellular structures. Life began as a chain of transmitted structures.

And here, communication became the decisive factor. Not the communication with speech and eyes, but deep: molecule to molecule, signal to receptor, form to function. Each complication represents a new way of communication.

Modern research shows that pre-cellular structures (e.g., proto-organelles) already had signaling mechanisms, where molecules react to changes in the environment: acidity, pressure, temperature. The beginnings of what we would later call metabolism, reception, even instinct – but at their core, there was always a signal and a response. Martin & Russell (2003): "On the origin of biochemistry at an alkaline hydrothermal vent," Nature Reviews Microbiology.

Life begins where the ability to distinguish arises from the reaction, where chemistry ceases to be random and becomes structured.

And the next evolutionary turn was the birth of the cell.

Few people think about it, but a cell is not just a building block of life. It is a highly organized ecosystem, a microscopic metropolis, where thousands of processes occur simultaneously: energy, protein production, logistics, waste disposal, command transmission and interaction with the outside world. Every second, billions of molecules follow strictly coordinated routes.

A cell is a structure capable of maintaining its role in a system. However, to participate fully as a whole, it requires not only DNA but also context, including signals, regulation, and coherence. The cell does not limit itself to executing a program; it recognizes its environment, adapts, and calibrates its reactions.

Hormones, neurotransmitters and electrical impulses form a complete system of coordination, without which the organism cannot exist for a moment. Communication is not an additional function of the cell, but a fundamental condition for its existence.

As organisms became more complex and multicellular, communication became critical: cells of different types had to act in concert. This is how the nervous system emerged: a mechanism for lightning-fast interaction.

The next step was the appearance of neurons. They do not build tissue, and do not digest food. Their only function is to connect. A turning point in biology has arrived: a fundamentally new biological role. Life oriented not to form, but to connection. Animal organisms develop a nervous system.

With the development of the nervous system, information began to circulate more quickly and accurately. Behavior as a response strategy appeared. Electrical impulses became a tool for quick decisions. The brain developed not as a controller, but as a node of connections.

And at some point, when the density of connections exceeded the threshold, the primary impulse of awareness arose. Communication gave birth to consciousness.

Consciousness has become a superstructure over the neural network: memory, prediction, symbols, language. We are aware of the act of communication itself, and therefore we can design it, strengthen it, and expand it.

First, the simplest nodes. Then, neural circuits appeared. Then distributed nerve centers. And finally, the brain: a superstructure capable of managing connections at all levels. This is not a storage of thoughts, but a dynamic coordination center — signaling, adaptive, and predictive.

Evolution is not the path of the strongest, but the path of those who synchronize better.

We are used to thinking that complex behavior necessarily requires a control center, and the brain is the main director of behavior. But if you look more closely, it becomes clear: the mechanisms of coordination and response to the environment work much earlier than the brain appears.

In some cases, there is no single control unit at all.

The connection is so fundamental that it can generate coordination, adaptation and even forms of strategy: without a center, without neurons, without consciousness.

Long before the brain appeared, life learned to be smart. Communication shaped behavior even in creatures without a nervous system.

Ants demonstrate a unique example of social organization. Their colony functions as a single mind, distributed among thousands of individuals. These insects create complex megastructures and respond to threats with astonishing precision, using only chemical signals, such as pheromones.

Ant colonies have an amazing collective memory. When one ant discovers a food source, the rest of the ant colony amplify its pheromone trail, creating a clear, directed route to the target. This communication system is so sophisticated that it allows the colony to make coordinated decisions, build complex architectural structures, and adapt to changing environmental conditions - all without any centralized control.

Scientific research confirms that the behavior of an ant colony is similar to the operation of a neural network (Gordon, 2010). Each individual acts as a carrier of an elementary signal, while the entire system as a whole demonstrates complex strategic behavior.

And ants are not the only example of network intelligence without centralized control. There are communication systems in nature that do without neurons at all, but still provide coordination, adaptation, and even mutual assistance.

Fungal mycelium forms an underground network that connects the roots of trees and plants in the so-called Wood Wide Web. Through this system, fungi transfer nutrients, warn of drought or pests, and sometimes restrict access to resources for plants that upset the balance of the ecosystem.

Research (Simard et al., 1997) describes this phenomenon as symbiotic communication, where fungi act as network operators for the entire ecosystem. Mycelium is the oldest form of communication that exists independently of the brain. This example proves that intelligent behavior does not necessarily require neurons, but can manifest itself as a property of a connected system. Complex communication processes are possible without a central brain - they arise where there is a dense network of connections and a need for coordinated actions.

Consciousness in this context becomes a superstructure, the foundation of which is always interaction. Mycelium demonstrates that the basic principles of intelligence are already embedded in the very nature of communication.

In other scientific works, researchers (Karban et al., 2014) have shown that plants also communicate through their roots, currents, and chemical signals. Communication is not an exclusive property of mammals. It is a universal biological mechanism available to all living things.

We see that intelligent behavior can arise without a brain, but it cannot arise without stable interaction. Fungi, plants, and insects all demonstrate that intelligence does not necessarily require a center, but it does require communication.

This is the key twist: intelligence is not always a product of neurons, but always a product of communication.

But it's not just nature that does this. Even in purely mathematical models, we see the same thing: a simple connection gives rise to a complex form.

Even without consciousness, without purpose, or a control center, if each element responds to the others, structure, stability, and even behavior can emerge. As a child, I loved playing tic-tac-toe with my friends. Simple moves to fill in the cells in a notebook with a cross or a zero created a strategy and led to victory.

Now imagine a world where no one controls anything. There is no mind, no plan, not even a goal. There is only a black and white grid - a field divided into cells. And each cell has a simple rule: if there are two or three living cells next to me, I continue to live. If there are fewer, I die of loneliness. If there are more, I perish from overload. If there are exactly three around, I come to life.

John Conway's Game of Life is a mathematical model based on cellular automata (Conway, 1970). It simulates a simple universe in which cells interact according to local laws, without a center or a conductor. Nevertheless, stable structures, cycles, and patterns appear in this universe. Some forms even move across the field, as if they had an intention, and build chains that are impossible to predict in advance.

How is this possible? How do such simple rules produce behavior that we perceive as complex, alive, and meaningful?

The answer is simple: from the connection. In the fact that each cell reacts to its environment. That each state is a consequence of the neighborhood. And the whole system is a field of interaction, in which the form is born not from the center, but from exchange. Without a program. Without an architect.

That's why this experiment so closely mirrors the principles by which reality works. Simple elements, linked together, create stability and cycles. Like neurons. Like societies. Like molecules. Like consciousness.

Consciousness is not an instruction manual, and it can't be downloaded. It grows through connection. And the more complex the connections are, the more complex the form becomes.

Evolution is not a selection of the strongest, but a selection of the connected. We see that the evolutionary path is not directed to the control center. It is a path to better tuning between the elements. The winners are not those who shout louder, but those who interact more precisely.

Consciousness appeared not when we learned to think, but when we learned to understand others.

Picture the African savannah a million years ago. A group of hominids spots the dark silhouettes of predators in the distance. But unlike other animals, these creatures do not run away in panic, nor do they rely on instinct alone. One of them gives a signal — a short, meaningful sound that the others instantly understand. They gather in a tight circle, pick up sticks and stones, and prepare to fight back.

At that moment, something happens in their brains that will set their species apart from thousands of others. They don't just react to danger – they transmit information, they strategize, they synchronize. This is no longer instinct. We see consciousness, enhanced by communication. And it is this that allowed humans to survive, and then begin the evolution of culture.

Observations and experiments with primates suggest that our closest relatives, chimpanzees and bonobos, already possess the art of systemic communication. However, they do not have a written language, let alone a civilization.

Chimpanzees can remember hundreds of words, construct simple phrases, and use language in everyday tasks. But what has become more important in recent years is what has been learned from observations in the wild.

Research at the Kokolopori Reserve in Congo (Berthet et al., Science, 2023) has shown that wild bonobos do more than just exchange signals. They combine individual sounds to change or clarify meaning. This is no longer just an alarm call or a summons. It is a real message, where one sound can complement or modify another, as in human speech.

Scientists recorded almost a thousand phrases and identified a set of 12 basic sounds that are combined into pairs. Bonobos use these combinations in different situations, and it is the context that determines how the message is understood. That is, the same words can mean different things depending on the circumstances, just like with us.

Similar observations in chimpanzees show that they transmit behavioral patterns, use tools, recognize their peers by voice, and even teach their young. **The connection goes beyond exchange.** It includes the ability to consider reactions and build meaningful interactions.

Communication gave birth to thinking. Thinking gave birth to civilization. Civilization is a network of interactions, exchanges, protocols that are supported in real time by billions of consciousnesses. But all this did not arise instantly, but the result of the accumulation of layers.

Each new layer arises as a natural consequence of the previous one. Just as in architecture, you can't build a dome without first constructing walls. And each wall holds only if the bricks support, align and build on themselves, the whole row, the wall and the whole structure. If there is a connection between them. This fundamental principle of interconnection is the manifestation of Communicational Energy.

But let's think about it - what exactly makes us unique? Why didn't even the most social animals build a civilization, but we did? How did the evolution of the brain turn communication into the main tool of survival?

The brain wasn't designed for communication. It was designed for survival. Its primary structures were responsible for fear, response to threats, and finding food. But at some point in evolution, a turning point occurred: it became easier to survive together.

Now let's take a broader view. Nature is full of examples where the complexity of communication becomes the basis for survival. Many animals living in groups have developed amazing ways of coordinating.

This is especially evident in wolves. A wolf pack is no longer just a group of predators, but a complex social system, where interaction occurs at all levels: from hierarchy to cooperation in hunting.

Wolves communicate through vocalizations (howls, growls, and squeals), body postures, facial expressions, and ritualized behaviors. These signals help assign roles, warn of threats, organize attacks, and maintain cohesion. Research shows that it is the coordination of actions, not the strength of an individual, that determines the success of a hunt.

While wolves communicate using visual and sound signals, other species communicate beyond the range of human perception.

Elephants are capable of transmitting messages over distances of up to 10 km using infrasound – frequencies below the threshold of human hearing. They use them to coordinate migrations, warn of predators, and call to watering holes. These echoes of a speech are a unique form of invisible dialogue between giants, reflecting the depth of their emotional and social consciousness (Poole, 2004). Elephant communication is a connection between memory, emotion, and strategy.

Humpback whales exhibit an amazing phenomenon of cultural transmission - their songs are not only preserved but also evolve from one generation to the next. In some cases, the entire population switches to a new melody in unison. Studies of the Pacific population (Noad et al., 2000) recorded cases where a new musical motif completely displaced the previous one in just a few months. This phenomenon confirms that communication processes can develop much faster than genetic evolution.

Honeybees use a complex system of symbolic communication, known as the dance. When returning to the hive, the bee uses specific movements to transmit the exact coordinates of the nectar source, including distance, direction, and even an assessment of the quality of the flower field (von Frisch, 1967). This system represents the first known example of spatial communication based on a rhythmic code.

Numerous examples, from fungal networks to whale communities, convincingly confirm the key idea: the evolutionary process involves not only struggle and adaptation, but also the improvement of the ability to communicate. The more accurately a biological system transmits, recognizes and interprets signals, the higher its adaptive potential and development prospects. Consciousness is a natural result of the evolutionary improvement of communication mechanisms.

Predicting the behavior of others is a huge survival advantage.

But even the most advanced forms of communication in the animal world were not enough. Yes, they could exchange signals and form behavioral patterns. But they did not create abstract concepts, build models, or convey complex ideas.

The real evolutionary leap occurred when our hominid ancestors began to develop the frontal lobes of the brain. They are responsible for abstract thinking, imagination, and the ability to predict the behavior of others. This gave Homo sapiens a fundamental advantage (Falk, 2004).

For thirty years now, the scientific community has believed that mirror neurons play a key role in this process. In the 1990s, Italian neurophysiologist Giacomo Rizzolatti made a discovery (Rizzolatti & Craighero, 2004) that shook up science. He found that primates have mirror neurons – special cells that are activated not only when the monkey does something itself, but also when it observes the actions of others. This meant that our brains literally reproduce the behavior of others within themselves. This became the basis for empathy, learning, and social interaction.

As society's complexity grew, so did the importance of effective communication. Those who sensed signals better found allies faster, coordinated better, and transferred knowledge more effectively, including to the next generation. It was no longer just the strongest who survived, but also the most connected. Not the fastest, but the most understanding.

An example of this is the fate of the Neanderthals.

Forty thousand years ago, there were two intelligent species on Earth: Homo sapiens and Neanderthals. Neanderthals were stronger; even their brains were bigger. Why did they disappear, and we remained? The main difference is communication. Neanderthals communicated with primitive sounds, and Sapiens had a developed language and could transmit complex knowledge; they built flexible social connections, and their communication allowed them to unite into larger groups.

Sapiens did not win because they were more perfect, but because they were better connected.

And perhaps evolution is not about dominance, but about connection, not about defeating others, but about the depth and precision of connection between ourselves.

But the path of communication does not end there. Today, we live in an era when consciousness is distributed among billions of points - people, devices, networks. Can a new phase of communication give birth to a new consciousness? How will this affect identity, culture, and perception of reality?

More about this in the following chapters.

Chapter 6 Summary:

- Communication is not just an exchange; it is a biological and evolutionary mechanism.
- Consciousness arose as a result of the growth of connectivity.
- Species evolution is the selection of those who communicate better.
- Civilization is a network of interactions that began with a gesture and ended with the Internet.

CHAPTER 7

NEUROBIOLOGY OF COMMUNICATION

Communication is not the transfer of
information. It is the transfer of attention
— Seth Godin

What happens inside us when we hear another person's voice? Why do some words stick in our minds for life, while others disappear? And why can one phrase change everything?

The answer is in the brain. But the brain is not just a signal filter. It is a system that creates meaning. It decides what is important, what is safe, and what is personal. It decides what is actually heard.

We have reached a key stage: in the previous chapter, we traced how communication accompanied the evolution of life, from amino acids to social animals. In the next one, we will look at how consciousness becomes a tool for overcoming. And here, at the center of this journey, is the mechanism that connects it all.

DOES THIS FEEL FAMILIAR?

Save an article about yoga on a social network, and the next day, your feed will offer meditation courses, mats, and retreats. Algorithms learn from us to suggest similars. They monitor what we respond to and adjust the reality around us.

But it's amazing how similar this is to how the brain works. When you think about something, it suddenly begins to seem that everything around is connected with it. There are signs, clues, and coincidences everywhere. There is a scientific explanation for this: selective attention and the Baader–Meinhof phenomenon. An effect in which, after the first mention of something, a person notices it everywhere. This is the result of attention + memory.

The brain, like an algorithm, enhances repetition and highlights the significant. We don't see everything, but we see what we are already tuned to.

We live in a world of endless information noise. Every second, the brain processes thousands of signals: advertising on the street, notifications on the phone, conversations around us. But we are not aware of 99% of them. Why?

Our brain filters information using three criteria:

- Emotions – anything associated with danger or joy – are remembered instantly. That is why news about disasters scares us, and joyful moments are imprinted in our memory forever.
- Novelty – the brain is designed to respond to the unexpected. Any non-standard phrase or image attracts attention (that's why marketers are always looking for unexpected moves).
- Personal significance – information that concerns us directly is remembered best. That is why a name spoken in a noisy room instantly stands out among hundreds of voices (the party effect).

But how did consciousness appear in living organisms in the first place? Consciousness is one of the most complex phenomena that still causes debate among scientists. How did it arise? Was it a gradual process that lasted for millions of years, or a sudden phenomenon that occurred at a certain point in evolution?

AND HOW DID COMMUNICATION INFLUENCE THIS PROCESS?

Today, **there are several key hypotheses about the origin of consciousness.**

1. The gradual complexity hypothesis (evolutionary model). This theory suggests that consciousness did not appear suddenly, but developed gradually, from simple forms to more complex ones. Reflexes → Instincts → Emotions → Abstract thinking. Early living organisms had only basic reflexes, but with the development of the nervous system, perceptual consciousness appeared - the ability to distinguish the environment and respond to changes.

2. Social Brain Theory (Robin Dunbar's hypothesis). This hypothesis states that consciousness emerged not for the sake of individual survival, but for the sake of effective communication in groups. The main idea: the larger the group, the more complex the social connections. To manage them, the brain had to analyze the intentions of others, build coalitions, and predict behavior. In essence, consciousness could have evolved not as a way of understanding the world, but as a mechanism for survival in complex social structures.

3. Theory of internal narrative (the talk to yourself model). According to this hypothesis (Carruthers, 2009), consciousness emerged when the brain began to conduct an internal dialogue. How did this happen? At first, people could only talk to others. Then the brain learned to talk to itself → reflexive consciousness emerged. This allowed us to build internal models of the world, plan, and fantasize. As evidence for this theory, they cite the fact that children under five often speak out loud, even when playing

alone. This may be a psychological trace of an ancient mechanism, when speech formed an internal dialogue.

4. Theory of language and symbolic thinking. Some scientists believe that consciousness could not have appeared before the emergence of complex language. Language = Self-awareness. The ability to think abstractly only appeared when words appeared to denote non-existent things. For example, the concept of the future is impossible without language - animals live in the moment, humans make plans and forecasts.

In addition to these evolutionary and psychological hypotheses, modern neuroscientists offer more formal theories based on brain observations:

- Global Neuronal Workspace Theory (Dehaene & Changeux, 2011)
  Consciousness is the result of the spread of information in the brain. When data becomes available to different modules (vision, memory, motor skills), it becomes conscious. This is the workspace where processes are coordinated.
- Integrated Information Theory (Tononi, 2004, 2016)
  Consciousness is related to the level of connectivity and complexity of a system. The higher the integration of information, the higher the level of awareness. This theory explains why even simple organisms can have a rudimentary form of awareness.
- Free Energy Principle / Predictive Coding (Friston, 2010)
  The brain is a predictive system. It constantly models reality by comparing expectations and sensory data. Errors trigger the correction of the model. Consciousness arises when this error becomes significant and requires updating the internal picture of the world.

All these theories agree on the main principle: consciousness is inextricably linked with communication. It is a mechanism for coordinating, predicting and transmitting information. Consciousness manifests itself in systems with a high density of interactions, where the need for complex coordination arises. The more intense and multifaceted the communication processes, the more pronounced the signs of conscious behavior.

It is here that we can formulate our hypothesis. Consciousness is not an object or a product of a single system. It is a node in the communication field. It appears where connections reach such a degree of complexity and density that the system begins to perceive itself as a whole. Consciousness is an effect arising from a structured connection.

There is also more direct evidence of the power of communication in shaping behavior.

There are documented cases of animals growing up outside their species and adopting behaviors typical of their alien environment.

A tiger cub raised among dogs began to display puppy-like behavior, asking for affection, wagging its tail, and reacting to signals that were alien to its instincts. A parrot living among cats imitated their behavior and even their way of interacting. Primates raised among humans adopted gestures and reactions that do not exist in the wild.

Innate instincts and reflexes all retreat before the force of the environment. The communication structure in which the organism is embedded begins to rewrite even what seemed to be fixed by nature.

The genetic program remains in them, but it is inferior. Because the communication environment forms habit, behavior, and perception.

It is not the species that determines the form of the mind, but the connection structure in which it is embedded.

An unusual but illustrative example is the octopus. This creature has exceptional intelligence, the ability to solve problems, recognize shapes, adapt, and even display character, despite not living in a society. At first glance, this contradicts our hypothesis: after all, intelligence, as we assume, arises through complex forms of interaction. But the octopus shows another side: **intelligence can be the result of deep internal communication.**

Up to two-thirds of its neurons are not in the brain, but in its tentacles. Each of its eight tentacles can autonomously make decisions, respond to changes, and seek a way out, all of which requires incredible coordination. In essence, the octopus's entire body is a fully formed distributed intelligence, where intelligent behavior is born from the need to coordinate actions between parts.

In this example, we see an important confirmation of the hypothesis: intelligence is formed not only through external communication processes but also due to complex mechanisms of internal synchronization. Communicational Energy manifests itself both as the ability to make contact with the environment and as the ability to organize connections within oneself. Intelligence arises where a structure arises, and a structure arises where a connection operates.

The next interesting example is one of the most mysterious forms of existence - a virus. It has no consciousness, no metabolism, and not even a cell. It is not considered alive in the classical sense. But it is capable of transmitting information, changing the genome, and triggering mutations. A virus is not an organism, but a message waiting to be unpacked. It does not develop on its own, but includes the possibility of developing within another who was able to accept it and understand it. It is a fragment of a code that becomes a material reality.

Biologists call it the boundary between the living and the nonliving.

From the standpoint of our hypothesis, a virus is a pure form of communication: it does not live for itself; it exists as an act of transmission.

And this is another confirmation: the idea of a communication information field is not an abstraction. If even in the material world there exist such quanta of communication as viruses, capable of changing living matter, then the very structure of interaction is fundamental.

And we are the result of such injections. About 8% of our genome is of viral origin. These ancient fragments are embedded in DNA as evolutionary messages.

Life develops not just as a series of random changes, but as a reaction to information challenges. It adapts, rewrites itself, and responds. Even on the border between the living and the nonliving, the principle of connection operates.

Consciousness is not something we have. It is how we connect.

The ability to establish connections is at the core of the very possibility of life. This principle remains true both for ancient amino acid interactions and for modern technological systems. Modern technologies represent new forms of Communicational Energy: artificial intelligence, global Internet networks, and digital interaction platforms. We exist in a period of unprecedented acceleration of communication processes, when the speed and complexity of information exchange reach a qualitatively new level.

The next stage of evolution is no longer just biology. It is a new level of interaction, emerging not only between people but also between man and machine, between biological consciousness and digital systems. However, the essence remains the same: connection, recognition, and meaning. Connection is a form of survival, and deep connection is becoming a condition for the future.

It can be assumed that evolution leads us not to dominance, but to dialogue. Not to struggle, but to agreement.

It is this ability to understand, synchronize and attune that makes us conscious beings. And perhaps even part of something greater.

If consciousness has always been linked to communication, the question arises: how will new technologies affect its evolution in the future?

In the following chapters, we will try to understand how the structure of communication and consciousness itself are changing in the digital environment.

The future of consciousness may be determined not by form, but by the quality of connections. The mind of the future may not be a brain, but a network. Not an accumulation, but a flow.

Consciousness is not a random flash, but a communication tool honed by the evolution of the Universe. It functions not as a passive recorder of events, but as an active creator of connections. This ability contains its fundamental essence and primary purpose.

⚙ COGNITIVE BLOCK - PRACTICE OF COHERENT PERCEPTION

1. Feel the movement of another
Mirror neurons are activated not by words, but by physical reflection.

❧ In a dialogue, gently repeat the rhythm of breathing, gesture, and the tilt of the interlocutor's head. Understanding begins with synchronicity. Tuning in with attention.

2. Slow down your reaction
Most responses are automatic. Mindfulness requires a pause.

❧ Before you react, mentally count to two. This switches the brain from a pattern to perception.

3. Set up your attention filter
❧ During the day, track what signals you respond to more readily: familiar, alarming, unexpected?

This will help you see how your internal algorithm works and begin to reconfigure it.

4. Notice what is repeated

◆ Notice when the same topic appears in different sources. This may not be a coincidence, but a manifestation of the focus effect. Attention can tune reality.

Chapter 7 Summary:

- Consciousness is the result of the interaction of information, communication and internal dialogue.
- We perceive the world through the filters of emotion, novelty and personal significance.
- Modern theories offer a neuroscientific explanation for the nature of consciousness.
- The concept of Communicational Energy describes the structure of communication that underlies learning, interaction and evolution.
- The future of consciousness is related to the quality of the connection, not the form of the carrier.
- A person is not only a participant, but also a creator of a new level of communication.

CHAPTER 8

GAMES OF CONSCIOUSNESS AND THE MOVEMENT OF MATTER

Everything we are is the result of our thoughts. With thought we create the world — Buddha (Dhammapada)

Just like me, you probably like watching sports competitions. Especially team sports such as football, hockey, or basketball. And, most likely, you have also noticed more than once: the favorite team, stronger in composition, statistics and even in the course of the match, leaves the field defeated. And the other, exhausted, unsure, seemingly already defeated, finds strength in itself and wins.

Is it fair or not? This question always sounds in the hearts of fans. In this chapter, we will try to find an answer to it. What is natural and necessary for the process of evolution of all living matter and consciousness in it? And why do we generally like to invent and participate in games?

We live at the intersection of matter and consciousness. Life is a journey through which consciousness interacts with matter. It tries, makes mistakes, adapts, and transforms. The DNA of any living being is not just a manual, but also an archive of adaptations, a trace of the evolutionary path, reflecting how consciousness, contained in matter, responded to the challenges of the environment.

Our DNA contains millions of mutations, and each mutation and change in the genetic code arises in response to external challenges. Mutations represent the imprinted memory of evolutionary development, embodied in biological form. This form has been honed through billions of attempts and millions of successful adaptations. Every cell, organ, and function has been transformed by the fundamental drive of living organisms to survive and overcome. Consciousness, at its core, has constantly sought ways to interact with the environment optimally.

The DNA code is all the realized desires and possibilities of organisms from the past, manifested in a specific present, on the way to their future.

Life is not a chaotic movement, but a directed aspiration. The aspiration is not just to survive, but to go beyond and become something more. From the first amino acids to man's flight into space, the entire history of life on Earth is permeated with the logic of overcoming.

There is an impulse within each of us that cannot be explained by instincts.

It is not limited to biology, survival, food or reproduction. It is an inner drive to understand and create, as well as adaptability and growth. This is the Communicational Energy of Consciousness, the essence of which is not in information, but in development, ascent, as well as in expanding horizons, oneself and together with others.

Anthropologist Michael Tomasello (2009, Why We Cooperate) conducted a series of studies comparing the behavior of young children and great apes. He discovered that the key difference between humans is the ability to act with a shared intention. Children, starting from an early age, strive not only to interact but also to coordinate actions with others for a common goal. They read intentions, predict their partner's steps, and coordinate their own without explicit instructions. And here we see not just a reaction, but a movement toward the future through connection.

This is what makes our species different: we can create a common future in our heads before it becomes reality. And we can convey this vision through gesture, word, and image.

Imagine: a man looks at the sky. He is small, almost insignificant against the backdrop of the Universe. But he does not lower his head - he raises it. His thought soars above the atmosphere, above the stars. He feels the scale, and instead of fear, delight appears. This is the nature of Consciousness - not to be afraid of heights, but to strive for them. Because there, at the top, the meaning becomes visible. Only by rising can you see the entire map of the material world around you. Admire it. And master it, at least in Consciousness. For hundreds of centuries, humans have been mastering new territories, waging hundreds of wars to master and rise above this world.

In cognitive psychology, this is described as a cognitive scale shift — the ability to see the system from above. It is directly related to the functions of the prefrontal cortex (Fuster, 2008).

Many philosophical and religious traditions since ancient times have also attempted to describe the connection between the thinking and the material. In the allegory of the cave, Plato depicted a man living among the shadows, but whose consciousness invariably strives for the light of truth. Descartes divided reality into a thinking substance (res cogitans) and an extended one (res extensa), emphasizing the duality of meaning and its bearer. Kabbalistic teaching described this connection through the metaphor of a vessel and the light that fills it.

All these images reflect one thing: consciousness does not simply observe matter. It strives to direct it. Communication between form and meaning becomes the main point of development.

Consciousness grows not in comfort, but in Overcoming

Why do people climb mountains? Have you ever thought about this question? It's cold, difficult, and dangerous there. But people go, again and again. Because the summit calls. Because somewhere inside us, there is a craving for heights, a craving to overcome. A mountain is a metaphor for all life. Each person has their own. Everyone has a choice: to stay at the foot or to climb.

This is how the game of Consciousness manifests itself. An eternal, subtle game: between limitations and possibilities, between fear and hope, between inertia and aspiration. And every time **you take a step forward, a new level of Consciousness is activated.**

But the ascent is not always heroic. Sometimes it is quiet. Internal. Overcoming fear, forgiveness, acceptance, effective adaptation, and creating something new - these are also steps up.

Life is a movement aimed at development, improvement and elevation. The reverse process - movement from better to worse - would deprive evolution of meaning and make life itself impossible. Matter moves in space, while consciousness makes its movement within matter. In essence, matter reflects the dynamics of consciousness.

The energy of communication, permeating our essence, requires the disclosure of potential. This disclosure is the essence of ascent - progressive movement forward and upward along the path of evolution.

Modern theories of consciousness, such as Embodied Cognition (Varela et al., 1991), argue that consciousness is inseparable from the body. It is born not in abstraction, but in action - in the process of interaction of the body with the surrounding world. Movement and perception become the basis of thought.

Modern theories of consciousness, such as Embodied Cognition (Varela et al., 1991), argue that consciousness is inseparable from the body. It is born not in abstraction, but in action - in the process of interaction of the body with the surrounding world. Movement and perception become the basis of thought.

Humanity has long tried to measure the level of consciousness. After all, if consciousness moves, it would be logical to find a way to measure this path.

From primitive exams to attempts to digitize intelligence in the IQ scale, we have looked for ways to measure it. But all of these are rather attempts to evaluate the result. In this book, we are interested in something else: not just measuring, but understanding the global mechanism: how reality tests consciousness. Not in the classroom, but in life. Not with a task, but with a situation. Not with an abstract scale, but with a real ability to cope, adapt, and lead matter through a challenge.

In the field of consciousness, progress is measured by the number of obstacles overcome and problems solved. Each successful overcoming increases the energy potential of the system, solving two problems creates a reserve of two units, while overcoming a thousand challenges means expanding opportunities by a thousand levels. Each obstacle becomes a step for development, and the accumulated experience forms the trajectory of the evolutionary path.

This principle is not a figurative comparison, but a working model of development. Consciousness evolves precisely through a collision with a certain boundary, with difficulties and the subsequent finding of ways to resolve them. The key aspects of this model are:

1. Quantitative measurability of progress
2. Direct dependence of potential on the number of solved problems
3. Practical applicability of the principle
4. No upper limit to development

But the tasks are different.

Some develop potential: strengthen, teach, and train. Others test for durability. These are challenges in which it is not growth that is at stake, but continuation itself. They become the real selection point for both an individual organism and a species.

Life is not a reward. It is like a challenge. Not a gift, but a process in which each being is tested for its ability to continue. Not just once, but every time. Evolution does not look at intentions and ambitions. It evaluates implementation. Hundreds of millions of sperm rush towards the goal, but only one reaches it. The rest are simply thanked for their participation.

In culture and nature, in history, it is not the potentially best who wins, but the one who embodies and brings the result. Who endured and worked precisely at the moment. And here, there is no place for the question of justice. We are observing the mechanics of life.

Every day of our lives, we take an exam. For resilience and adaptability. For internal flexibility and the ability to cope with a new task. You were building up mass, now you need lightness. You were walking in a straight line, but now the path goes sideways. It is not just the strong who survive, but the ones who are sensitive to change.

In essence, all living things exist in two modes: accumulation of potential and its testing. Antelopes grazing, students at a lecture, or chicks in a nest accumulate energy, knowledge, and skills. But which of them will continue on their way, the crisis will show. It could be a lion in the savannah or an exam in a session.

And it's not about the volume of information, but about fine-tuning. Who will notice the threat first? Who will choose the right strategy?

Consciousness that is capable of prediction, of an accurate reaction to the context of a situation, is the consciousness that leads matter through a crisis.

A crisis is not an accident or a failure, but a moment of truth. It does not simply test – it selects. It cuts off the excessive and preserves the stable and viable. Any crisis is so important for the development of each material individual and is important for the evolution of the species as a whole. The universal criterion of selection is the ability to change while preserving the inner core.

You, just like me, probably often felt indignant: where do these problems come from? Why did the crisis come now? Everything was so good...

Over the years of my entrepreneurial life, I have co-owned about a dozen businesses. Some of them were very successful, others were painfully instructive. But they all had one thing in common: crises came everywhere.

At first, there is always shock. Then chaos, dozens of decisions a day, a fight for time. And if you are lucky enough to get through this, then the main thing appears: an understanding of how to prepare for the next one.

One of the toughest episodes in my life was the crisis in the agricultural enterprise to which my brother and I had dedicated more than 12 years. At that time, more than 350 people worked there. When the market collapsed, nearly a third of the companies in the industry failed to survive. And for us, the days began in which every day was like the last chance.

Every morning I woke up with one thought: how to survive another day. How and where to find more reserves. How to convince suppliers that we will pay off. How to maintain trust. How to prevent the system from collapsing. Calls. Decisions. Debts that seemed impossible to ever pay off. But life tests you in this way. The crisis takes away survivability and clears the space for the actions of the survivors. Less than 8 months passed, and we recovered and grew, and paid off all creditors and suppliers in full. Not because we knew in advance. But because we were flexible, focused, and engaged. The crisis cleared the field. It did not forgive mistakes, but gave the opportunity to survive to those who were ready to change.

This is not unique. In big business, such examples often end quite differently.

In 2007, Nokia controlled almost half of the mobile phone market (50.8% of the market, about 150 billion euros in company capitalization). It seemed unshakable.

But that same year, Apple presented the iPhone. And nothing happened at Nokia.

No reaction. No abrupt restructuring. No internal panic.

Their loss was not in the product. It was in the inertia.

Nokia missed a signal (just as a phone notification - you have two missed unanswered calls).

And a few years later, the company, once worth tens of billions, was sold for next to nothing (in 2014 for 5.4 billion euros, of which 30% was the cost of patents and licenses).

Such stories are called the fate of the dinosaurs. They dominated the Earth's ecosystem 65 billion years ago. But when the crisis arrived, it was not the largest and strongest that survived. But the most adaptive.

Therefore, in our understanding, Life is a series of challenges, where a specific matter is tested for survival. And for Consciousness, contained in this matter, there appears an opportunity for development and complication, the development of a path to the future. This is how complexity develops. This is how form is honed. This is how evolution becomes not just biology, but the logic of communication.

The theory of Mental Effort & Cognitive Load (Sweller, 1998) explains: consciousness grows through the complexity of tasks. The higher the internal load when solving problems, the greater the growth. From this, we can derive a formula where **Functional load = a measure of growth and evolutionary advancement.**

The entire DNA record, all changes in matter, all mutations and habits, all reflexes and abilities – all this is the contribution of Consciousness to the future of matter. In neuropsychology, it is confirmed: the anterior cingulate cortex is activated in moments of cognitive conflict and difficult choice. This part of the brain helps choose a path and learn through overcoming (Botvinick et al., 2001).

Consciousness is not a given, but a path. It cannot simply be had – it can only be built. Every step toward greater understanding, greater overcoming, and greater responsibility becomes a step up. And this ascent is not about becoming better than others. It means becoming more than yourself.

I have seen many people who sign up for dozens of trainings, read hundreds of books, and sincerely believe that this is enough to change their lives. Of course, knowledge is important and is the foundation for making decisions. But without tuning into reality, without transforming knowledge into action, they are just a burden, not a leap forward.

Consciousness grows not from reading, but from choosing, not from plans, but from attempts. It opens up when it encounters difficulty and does not retreat.

Consciousness grows when it resists, when it asks questions that have not yet been asked. When it chooses not the shortest path, but the path on which one can learn. The real movement of Consciousness is not a process of running away from pain, but a movement through it.

As Friedrich Nietzsche wrote: What does not kill us makes us stronger. But let us add: it makes us deeper. More complex.

Higher. Because every wound that we become aware of becomes a step.

And this is the path to gaining inner independence and stability. When external circumstances collapse, but you still move forward. When you are no longer an object of reactions, but a subject of meanings. When you choose what your victories and defeats mean to you.

The higher the ability of consciousness to precisely interact with matter, the more stable the system. And the higher the chance of surviving the next crisis, to move on to a new stage.

Crisis is a mirror in which Consciousness sees its readiness for the future

---

Consciousness is not just the ability to perceive. It is also the ability to collect, direct and redefine from a simple reflex to instinct, from instinct to prediction, from prediction to strategy and meaning. With each turn of evolution, the density of connections increased. Consciousness became denser, more complex, embracing the world around it ever more deeply.

Humans are the only ones who have gone beyond adaptation. We not only adapt, but also project: alternative realities, new forms, future states. This makes us carriers of a special kind of Communicational Energy. We not only perceive it, but also form it, amplify it, and retransmit it. We are active links in this structure.

Consciousness is a system, and like any system, it has a structure. At any point in development, three levels can be distinguished:

1. The common is the field of consciousness in which we are immersed: the collective mind, culture, the laws of nature, and other consciousnesses.
2. Individual is your unique point of consciousness, anchored in the body, in biography, in context.

3. Communicational Energy is the connecting thread between the first and the second. It is the force that allows us to connect the external and the internal, to feel and change, to understand and act.

Everything alive is always connected. We exist in communication. Thoughts, emotions, and choices are not personal characteristics, but acts of coordination with a larger field.

This structure is universal and recognizable. It attempts to explain how the personal and the collective become a system, not just a collection of random points.

But there is one more important thing. We all live and act within a communication field. With our desires, decisions, and will, we send waves into this field. They are reflected, collide, amplify, or fade away. We shape reality – but not alone.

Each of us encounters other people's «waves». Other people's projections, vectors, and actions. They can coincide or confuse. Sometimes they support, sometimes they reject. Therefore, a natural question arises: where am I going? Am I on my own path? What am I guided by?

In the search for such reference points, the topic of objectivity arises. Not as a dogma, but as a way of verification. How to feel that my movement is not just a reaction, but a realization. That my desires are not imposed, but objective and genuine.

And here we come to the main point: there are only three things that can serve as navigation within the flow:

- **Entry point** — the moment when you appeared as consciousness. This is the starting coordinate.
- **Stream** of reality — movement of the general communication field, including all laws and interactions. Something that cannot be stopped, but can be used.
- **Your trajectory** — the line you build with your choices, your effort, your attention. The trail you make by interacting with this flow.

Everything else is variable and subjective. Thoughts, opinions, roles, and even emotions are important, but not constant. They are signals, not coordinates.

Only movement is true. Your path of interaction. And true stability is not in a frozen position, but in the ability to move without losing touch with yourself, where your actions correspond to inner clarity, and not external circumstances.

Communicational Energy allows signals and information to be transmitted in this world, and also to create structures from chaos. But the main thing is that it allows one to improve and develop, through desires, movements and overcoming.

This energy fills the space with meaning. A person in this process is not an observer, but an active participant. Where every thought and action becomes an opportunity to change not only yourself, but also the reality around you, this is the main rule of the evolutionary game and the victory in it.

COGNITIVE BLOCK: CONSCIOUSNESS IN CRISIS

Exercise 1. Analysis of previous crises
• Think back to 1-2 crises in your life or work that became turning points.
• What seemed impossible back then but became your strength?

Exercise 2. Growth points and crisis forecast
• Where are the main growth points in your project or personal communication system now?
• What crisis situations may arise in the future, when (in what perspective), and under what conditions?

Exercise 3. Crisis preparedness
• Imagine: tomorrow there is a crisis (a partner's refusal, a failure of trust, a challenge from the environment). What in your system and in you can withstand the pressure? What might not withstand it - and why?
• Where is the real center of your stability?

Exercise 4. Anticipation and reaction

• In what areas of your life/work does consciousness work as foresight?
• Where does it act only as a reaction?
• What can you change now to be proactive rather than reactive?

Chapter 8 Summary:

- Consciousness and matter are inseparable: matter is the carrier, and consciousness is the driving force that forms and comprehends it.
- Life is a journey in which consciousness overcomes challenges through matter, expanding the potential for interaction and adaptation.
- Development is not linear. It occurs through the accumulation of potential and the passage of crises, where selection is not based on strength, but on the accuracy of conscious adjustment.
- Consciousness grows through overcoming, acquiring form and meaning in resistance, in effort, in choosing the non-obvious.
- A person is a unique generator of Communicational Energy, capable not only of adapting, but also of creating new meanings, models and realities.

CHAPTER 9

REALITY CODE. FROM ZERO TO ONE.

*At the core of reality is the bit - a
quantum of information that cannot be
divided — John Archibald Wheeler*

Once, at an international scientific symposium attended by several
thousand mathematicians and after a series of intense reports,
everyone went to the nearest bar. The first mathematician ordered
one glass of beer, the second, half, the third, a quarter of a glass,
the fourth, an eighth of a glass. At the fifth mathematician, the
wise bartender, understanding the logic of the orders, interrupted
them with a smile: Here are two glasses of beer for you. Enough
for all the participants of the symposium.

After these words, there was no drunken brawl or philosophical
discussion. Not because the scientists were too sober or polite.
But because this is not a real story, but a mathematical joke. It has
everything: paradox, humor, and mathematical laws. But the main
thing is the observation and description of reality.

Since its inception, humanity has tried to make sense of this
world. First through numbers. Counting, proportions, and
repetitions. Then, through calculations and codes. And as we

evolved, this logic only grew stronger: we increasingly sought to explain reality through digital laws. Cyclic. Algorithmic. Structural.

In all cultures, religions and philosophical systems, man has tried to find a universal principle for the structure of the world, and almost always it was the principle of number. From Egyptian arithmetic to Kabbalistic texts, from Euclid to modern cryptography - everything is built on a digital communication code, in which not only is information transmitted, but a structure is created.

In the previous chapters, we outlined our key hypothesis: life and consciousness arise not simply in matter, but in a field of interactions—dynamic, algorithmic, deeply connected to information. We called this field Communicational Energy.

But what exactly is happening in this field? What principles and levels exist? How is order formed, moving from the potential to the manifest?

If Communicational Energy can be the basis for the formation of life, matter and meaning, then a natural question arises: is it possible to somehow describe this interaction? Measure it? Model it?

This chapter aims to examine the structure of interaction. We will address the number not as an abstract symbol, but as a form of presence. To the number as a code in which not only digital systems but also the laws of reality itself unfold.

Modern science, at the intersection of physics, mathematics and computer science, already records repeating patterns, sees stable models and introduces terms like computable symmetry, information noise, and algorithmic entropy. These concepts reflect the desire to describe the world through codes, structure and computability. But science is not yet able to explain why exactly such patterns underlie the universe. And here the space of hypotheses begins.

An interesting example is the story of Mitchell Feigenbaum, a physicist who, in the 1970s, while studying chaos, discovered in it... a digital pattern. He noticed that the most diverse physical systems, from water drops to electric current oscillations, when moving to chaos, obey the same number: 4.669.... Today, it is known as the Feigenbaum constant, like the signature of an algorithm, built into nature. We record repetition, but do not understand the mechanism. It turned out that chaos is also a form of order, just hidden. Even where it seems that nothing repeats, there is an unfolding code.

If a number can describe even chaos, perhaps it is the bearer of meaning. Then it is not the form that generates logic, but logic that shapes the form. And then the idea becomes clear: reality is an algorithm of interaction. And the number is its first letter.

We do not claim strict scientific proof. We understand that technology does not yet allow us to verify all the assumptions in this chapter. But perhaps this attempt will become a contribution to the general map of knowledge. A bridge between physics and philosophy, mathematics and meaning.

And if we continue the theme of mathematical humor, then for this case, mathematicians have the following joke prepared:

— What do you get when you cross a mathematician and God?
— The result will be a God who cannot prove his existence.

We are used to imagining scientists as people in offices, surrounded by formulas and graphs. Meanwhile, one of the oldest philosophical debates continues to this day: is mathematics an invention or a discovery?

Imagine mathematicians not at their desks, but as Columbus or Magellan – explorers heading into the unknown without maps or guarantees, guided only by intuition and basic knowledge. And suddenly, an unknown shore appears on the horizon: an information continent where numbers are not an abstraction, but the structure of being.

Mathematicians become not inventors of formulas, but discoverers of an eternal continent that existed before them. Each new step is like a flag on a fixed territory of understanding.

Max Tegmark, a professor at the Massachusetts Institute of Technology (MIT), has hypothesized: We don't just use mathematics to describe the world – we live inside a mathematical object.

And Eugene Wigner, one of the founders of quantum physics, wrote: The incredible effectiveness of mathematics in the natural sciences is a miracle that science itself cannot explain.

Mathematics is not just a calculation tool and a school subject. It is a universal language of connection and interaction, in which the Universe records its laws, its existence and meaning.

And all that we are only beginning to understand is only the first coordinates on the edge of a new continent: the informational communication field of existence, where dozens of expeditions and, perhaps, even great discoveries await us.

Everything is number (Greek: Τὰ πάντα ἀριθμός ἐστι - Ta panta arithmos esti), said the ancient Greek philosopher and mathematician Pythagoras, several thousand years before the advent of programming. He believed that the structure of the universe could be expressed in numerical relationships and that numbers were a form of higher harmony.

Pythagoras believed that numbers were the basis of everything that exists, and that the entire structure of the universe could be expressed through mathematical relationships and proportions. This idea had a profound influence not only on mathematics but also on philosophy, music, architecture, and even mystical teachings.

Modern science does not deny the numerical nature of the world. On the contrary, in physics, computer science, biology and cosmology, numbers have become a universal way to describe the structure of reality, from quantum states and symmetries to genetic code and blockchain algorithms.

Srinivasa Ramanujan (1887-1920) is a name that still inspires awe in mathematicians. This young man from India, without any formal education, created more than four thousand formulas, many of which remain a mystery to this day. Scientists admit that his discoveries are comparable in depth and originality to the works of the greatest European minds, from Euler to Gauss.

He believed that numbers are not just tools, but the language of the Universe. And he once said, Equality has no meaning unless it expresses the thought of God.

Ramanujan claimed that he received his discoveries from the goddess Namagiri in a dream, as revelations. And, surprisingly, many of these formulas were confirmed as fundamental decades later. This is evidence that numbers can be more than quantity. Numbers can be the thinking structure of the Universe, a bridge between spirit and form, a formula in which the breath of the universe sounds. Numbers as a spiritualized essence. Formula as an expression of meaning.

Like Pythagoras, we will try to draw analogies between the digital series from 0 to 10 and its possible semantic content. **Each number is not just a symbol, but a level, a code, a proportion of the reflection of the Universe.**

Similar ideas were reflected in religious and mystical systems. In Kabbalah, the 10 Sephiroth describe the stages of the descent of Divine Light into matter, each with a certain energetic and semantic characteristic. The stages through which the Almighty creates, controls and maintains the world. This is the structure of spiritual reality, a kind of map of the descent of the Incomprehensible to our material level. This is, in essence, an algorithm of communication between the absolute and the manifested.

The Sephiroth are not things or forces but principles, stages and qualities. They can be thought of as 10 rays or filters through which the energy of creation flows.

The intuitive sense of the numerical structure of the world remained in the realm of metaphysics for a long time. But as science developed, the same principles began to be sought by rigorous methods. Humanity not only believed in the meaning of numbers, but it also sought to verify them.

Scientific knowledge is based on verification. Any hypothesis is tested by formulas, relationships, and calculations. This is the algorithmic language of Consciousness, which seeks to describe and comprehend reality.

Paradoxically, even when creating virtual, artificial levels of reality – from computer models to digital worlds – we again rely on the same numbers, simply reduced to 0 and 1. Everything we learn, analyze or recreate is built on numerical systems and proportions. In any process where structure arises, a number also appears as a form of expressing order. Perhaps this is the universal language of interaction: algorithmic logic, in which not only thinking but also reality itself unfolds.

Our task is to try to see in each number not just a quantitative sign, but a level, a section, a proportion of the reflection of the structure of the world. Number as a structural element of the laws by which any ordered process unfolds: from physics to logic, from biology to information.

The number 0 and the concept of zero. The beginning, emptiness, everything.

Europe adopted the concept of zero relatively recently, only in the Middle Ages, while India, China and the Arab world used zero in calculations much earlier, perhaps because the human brain is accustomed to paying attention to and describing only what is there.

Historically, we have become accustomed to calling decimal digits Arabic, but in fact, their origins are Indian. It was the Indian mathematician Brahmagupta who introduced the symbol for zero, suniya, in the 7th century. This word meant not just nothing, but also pause, space, and sky or even universal emptiness. In the

Indian tradition, numbers were transmitted orally — numbers were spoken as sequences, and a pause was required to separate one digit from another. Thus, emptiness acquired a mathematical meaning: not as an absence, but as an interval that structures speech and thought.

Adding zero to a number did not destroy it, but increased its order. This is what gave mathematics the path to infinity: any scale can be expressed through 0. Zero is the keeper of any scale and the operator of infinite ordinal growth.

You've probably often heard the phrase, How many zeros are there in his account, a phrase that is often heard in films and daily business. But it's not just about the number of digits. It's a question of level. A question of scale. And the difference is sometimes colossal. One of the most striking examples: a million seconds is only 12 days ago. And a billion seconds? That's already 33 years. That's what a difference of a few zeros means.

Zero is not just a number, but a transition value, a boundary between positive and negative, a window into another order. It serves as a neutral element of addition, becoming the basis of the positional notation system - the principle by which the value of a digit depends on its place in the number. This is what allowed humanity to operate with large numbers, introduce digits and build the entire digital era. Thanks to zero, mathematics acquired not only precision, but also scalability, and with it, all modern science, philosophy and technology.

Zero is both emptiness and potential. The point where everything is already possible.

In philosophical and religious systems, zero has always been more than just a number. In Eastern philosophy, it is shunyata, the void from which everything is born. In Kabbalah, it is Kether, the beginning, where everything is not yet manifested, but already exists as a plan.

Modern physics is also rewriting the meaning of emptiness. Vacuum is not an absence, but a quantum foam. A field where virtual particles flash and disappear, confirmed by the Casimir and Lamb effects. According to the Casimir effect, even in empty space, measurable forces arise as a result of quantum fluctuations. Emptiness is not empty: it is potential.

Zero does not represent absence, but a state of complete readiness. It is not nothing, but everything that can be. Zero is not a number in the usual sense—it is a starting point from which numbers can emerge. It is the pause that precedes any movement.

If a number is a structure, then zero creates the conditions for its appearance. It marks the boundary between the potential and the manifested. Zero itself does not exist - it manifests its reality only through interaction. Zero acquires meaning exclusively against the background of everything that exists - as a frame for a picture, as a mirror for reflection, as emptiness without which it is impossible to recognize fullness.

In the context of our hypothesis of Communicational Energy, about the existence of an algorithmic dynamic communication field, zero is the very possibility of its existence and its real existence at the same time. Zero is not the opposite of something, but its background, its contour. It is in this background that the communication field begins to work: it outlines, connects, and fills. Everything that can manifest itself first appears as a possibility, and this is zero.

If we try to comprehend the essence of zero even more deeply, then 0 is a perfect symbol, containing the paradox of existence: it is simultaneously empty and full, nothing and everything. If the Universe is a cycle, an eternal circulation of energy and matter, then zero is its ideal expression, the boundary between potential and manifestation. Our Consciousness, the Communicational Energy that creates and fills it, is not a point in this circle, but the circular strip itself, the boundary, the outline of which already in the process itself separates the manifested from the unmanifested.

The communicative energy of Consciousness cannot exist without the potential for its manifestation, just as empty space does not acquire meaning until Consciousness shapes it into a meaningful form. In this interaction, the material world is born, in which law, logos, and awareness become fundamental properties of matter.

This «Eternal dance of zero» is the dynamics of creation in the Universe. Consciousness (the point of interaction in the algorithmic field) outlines the boundaries, separating something from nothing, turning chaos into order, potential into manifestation. But this process is also cyclical: any manifestation can return to nothing, dissolve in limitless potential, and then be born again in a new quality.

Thus, zero is not just an absence, but a state of dynamic equilibrium between potential and realization, between emptiness and content. It represents the starting point where opposites converge to form the basis of structure: level by level, meaning by meaning, number by number.

Zero is not frozen everything, but eternity in cyclical motion, just like a wheel: empty in the center, but generating momentum through contact with the surface. The infinite is realized at a specific point of contact. Zero functions as a condition of movement, like a central axis in a mechanism that does not move itself, but allows the entire system to rotate. It is in this emptiness that momentum arises: interaction becomes possible, and concrete reality is formed from abstract potentiality.

This is how life emerges, moment by moment. This is in line with the ideas of Eastern philosophy, where the true essence of being is not found in stillness, but in conscious contact with the present moment. Consciousness does not simply record reality, but interacts with it, setting the rhythm of the unfolding order. In this cyclical process, zero acts as the primary trigger point, formal and energetic.

THE FIRST STAGE. UNIT. THE ACT OF EMERGENCE. THE START OF MOVEMENT AND ORDER.

In the beginning, there was a beginning. Something extraordinary. Something that did not exist before, at least in these coordinates of the world, in these axes of reference. A great impulse, the launch of a new round of being, a new cycle of existence. A reversal of proportions. A new starting point. It was the birth of something in something. The very essence of emergence.

Today, the dominant scientific theory of the origin of our universe is the Big Bang. It literally describes the extreme expansion of space from a superdense state. Indeed, it is supported by several observational facts: the removal of galaxies, the uniform microwave background, and the distribution of the primary elements.

But if we take this not literally, but as a flash of transition, then perhaps another name would be more appropriate: the Big Switch-On.

It was as if someone had flicked a switch, and the first wave emerged from the silence.

It was as if someone had returned to a cold house and filled it with warmth and movement.

Just as we turn on a kettle and the water inside begins to bubble and heat up, or press a button on the remote and the screen comes to life, turning pixels into a picture.

The same is here: one is not just a number, but the first impulse of meaning. The first level of interaction. The initial spark in the algorithmic field that fills everything and from which everything arises.

The unit can be called the first facet of order, the starting point and the symbol of existence. In mathematics, it plays a key role as the neutral element of multiplication ($1 \times x = x$), the basis of the number system and the smallest indivisible quantity. If zero is potentiality, emptiness ready to be filled, then the unit is the act of emergence, the first structure in chaos.

One has no internal complexity, but it sets the direction. This number is the starting point. Without it, neither two nor a thousand can be distinguished. It is the first choice. The first step. The first vector.

In philosophy, the unit represents integrity, individuality, and self-sufficiency. In the Platonic tradition, it is associated with the idea of the One, the basis of all things, from which numbers and forms are born. In Eastern philosophies, such as Taoism, the unit symbolizes harmony, the transition from emptiness to being, and in religious systems, the highest essence, the root cause of the universe.

In the scientific sense, a unit often means a standard, a measurement standard, or a fundamental constant. In quantum mechanics, it can be associated with the discreteness of being, elementary particles that cannot be divided without losing meaning.

The process of development began. Not just a line, but a comprehensive expansion, and accordingly a comprehensive development. A point of origin appeared, and from it the first wave of filling unfolded. In this impulse is the further birth of matter, the birth of direction, the birth of meaning. An explosion not as destruction, but as formation: as the first act of inclusion, setting the vector.

The number one is not just the first number, but the fundamental starting structure. It is where the movement begins. It contains the initial impulse, containing all the potential for the subsequent development of the algorithm.

The Unit is the Consciousness that permeates the entire surrounding world.

This number expresses the point of filling the world, the state of perception, knowledge and reflection of the entire world. It embodies movement - coming from zero, returning to zero, existing in cycles. In this movement, tens of zeros, twos, threes and complex numerical sequences are born. However, in the structure of any number, one always remains.

The number is not what counts. The number is what connects.

The famous theoretical physicist, one of the creators of string theory, Leonard Susskind, proposed the holographic principle: everything that we perceive as a three-dimensional Universe can be a projection of information encoded on the two-dimensional boundary of space. That is, all of reality can be a hologram - not a volumetric entity, but a reflection of patterns, interactions and numbers distributed on an invisible plane.

But if we look deeper — and connect this with our hypothesis — then perhaps we live inside a dynamic hologram. Not a mirror of form, but a field of movement. We do not simply see a projection of an object, but observe traces of its impulses — energetic interactions, flashes of communication, microwaves of algorithmic activity. In this case, a hologram is not a photograph, but a flow. A reflection not of a static object, but of an energy field that pulsates, communicates, and structures itself.

Then the number ceases to be just a measure. It becomes a carrier of being, an algorithm for the unfolding of form in time, a language with which the field and perhaps the Universe itself talks to itself through the mirror of Consciousness.

This is what Pythagoras felt when he said, *Everything is number.* And this is probably what Ramanujan perceived when he received his formulas as revelations. He did not invent numbers - he heard them as a performer hears music when he reads notes from a score.

The holographic principle can also explain the nature of Communicational Energy: it does not move from point A to point B, but is reflected throughout the entire system, like a holographic signal - instantly, at every point.

This is **the property of numbers - to be both local and universal at the same time. To be information and form.** This is how the digital alchemy of the Universe appears - where numbers do not describe the world, but shape it at the moment of interaction.

Numbers are instruments of measurement. But they are also carriers of meanings, according to which the world unfolds into form. Numbers become form, and form in turn becomes a way of the evolution of consciousness. And therefore, we have grounds to consider evolution as a communication algorithm, unfolding in time as a sequence of meanings, encrypted in numbers.

And perhaps this intuition was embedded from the very beginning - even in the drawing of the numbers themselves. Already in their visual form, ancient people seemed to guess: counting is not just a quantity, it is a path in which each subsequent unit grows from the previous one, continuing its impulse.

In the earliest Indian numerical systems, the symbols of numbers were not strictly distinguishable signs, but were built on an internal logic of continuity. Visually, they were similar to each other, as if they were born from each other, indicating an internal connection and development. The external similarity of neighboring numbers and the internal logic of their design indicated that numbers were understood as parts of a single flow.

The very foundations of counting already contained the idea of communication: not as a technical recount, but as an internal movement in which each unit continues the previous one, not interrupting, but developing it. Even before the emergence of information theories and algorithms, the ancient Indians laid the very principle of connecting energy in counting.

In this chapter, we tried to assume and understand how the number became a mirror of evolution, and evolution an algorithm of communication, that the number is the rhythm of existence. And to understand the interactions by which everything lives and returns.

Because in the structure of the universe, like in a mirror, the same archetypes are reflected again and again: birth from emptiness, differentiation, movement, balance, overcoming and returning to the source. And we will talk about this in the next chapter. We will move on to the numbers that form structure, difference and direction - 2, 3, 4 and 5.

Chapter 9 Summary:

- Numbers are not just symbols, but universal codes of structure and meaning of existence.
- Zero is not an absence, but a potential. It opened the way to infinite order and became a symbol of readiness and beginning.
- The unit is the act of manifestation. The first impulse from which all the diversity of forms unfolds.
- The numerical system carries a deep logic of communication. Even the visual continuity of numbers is an intuitive model of the communication flow.

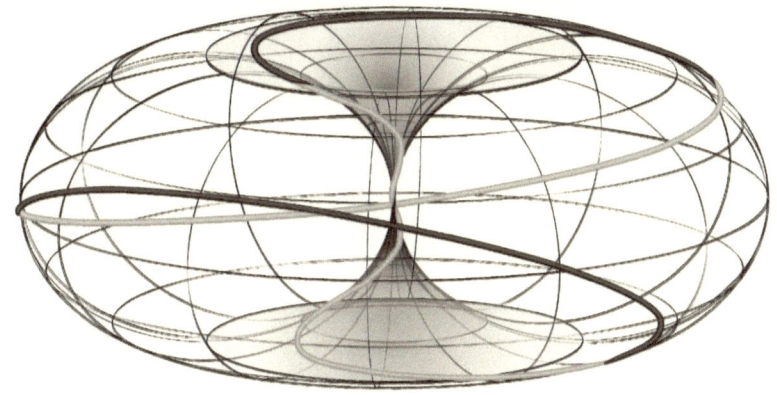

The Toroidal Structure of Reality

One of the most elegant and dynamic geometries in physics, the **torus** is considered a candidate for describing a **finite yet unbounded Universe** — a shape in which energy, matter, and information can circulate endlessly without a beginning or an edge.

The torus represents a **self-regulating, self-sustaining field**. Its topology supports continuous flow, feedback, and return — qualities aligned with both **universal energy dynamics** and **systems thinking**.

In theoretical physics, some cosmological models propose that the Universe may have a toroidal shape, allowing for both **infinite expansion** and **structural closure** without singularity.

As a visual metaphor, the torus also reflects the logic of **digital energy levels** — where "0" and "1" are not static positions, but dynamic poles in a rotating field of potential. Meaning arises not from fixed values, but from **curvature, circulation, and interaction** within the field.

CHAPTER 10

DIGITAL ALCHEMY (2-5): TRANSITION FROM DIFFERENCE TO STRUCTURE

The universe is written in the language of mathematics, and its symbols are triangles, circles, and other geometric figures — Galileo Galilei

We live in a world where any number is constructed from just ten symbols: from 0 to 9.

Everything we know about counting, math, and coding unfolds on this foundation.

Have you ever wondered - but why exactly ten? Why not 12, not 17, not 64? Why does the cycle end with nine and start again from zero?

The first answer is that the main reason is anthropological. A person has ten fingers, and since ancient times, counting has been done on them. The decimal system seems natural because it is literally built into our body.

But imagine: if we had 12 fingers or only 8, would we have different mathematics? Different equations? Different laws?

History knows alternatives: the Babylonians used a sexagesimal system, the Mayans a 20-based system. In England, up until the 19th century, monetary and measuring units were subordinated to the dozen.

Modern digital technologies also rely on binary or hexadecimal notation.

However, it was the decimal system that became universal: in economics, science, education, and computing.

And here an important point arises: human counting went from 1 to 10, but the modern system goes from 0 to 9.

Zero is not expressed by fingers; it cannot be counted on the hand. It appeared not from the body, but from the mind. First as emptiness, then as a symbol, then as a full-fledged number.

It was the emergence of zero that gave humanity a positional number system, flexible, logical, and expandable. It was a mathematical revolution.

Ten digits have become a closed system. Moreover, this system is universal.

No matter what number system we work in, binary, hexadecimal or decimal, the meaning of numbers remains.

The number 29 in decimal, 0x1D for a programmer, and a symbol in the Mayan system, all of these are the same numerical state, represented by different alphabets.

Mathematical laws do not change:

- $2 + 2 = 4$ in any system,
- multiplicity and divisibility,
- transitions between digits,

- proportions and symmetry.

This is what tells us the main thing: the logic of numbers is universal.

This means that the fact that all numbers can be expressed through 10 basic symbols is perhaps not a coincidence, but a reflection of a universal principle of interaction.

Just as the periodic table describes the laws of matter, just as the spectrum of light covers everything visible, so the numbers from 0 to 9 perhaps reflect the atoms of counting.

And although different systems use different bases (2, 8, 16...), only in the decimal system this base is closed from 0 to 1 in the decimal fraction: 0.1, 0.2, ... 0.9, 1.0 - symmetrically and without remainder.

The entire mathematical structure of the Universe can be built from these ten digits. This is not only a convenience of notation, but a reflection of the very idea of development and transition: when the digits run out, the system adds a digit and opens a new level.

This is the interaction algorithm. From zero to one. From simple to complex. Through repetition and increasing difference.

This is how communication logic works: not an abstract formula, but a model in which the world holds, reproduces and comprehends itself.

Perhaps this is why the decimal system became the main one. Not because of the fingers. But because it perfectly realizes symmetry, completeness and closure of the cycle.

Numbers are not just a measure. They are events. Energy states. So 0 and 1 are not just numbers. They are the beginning of a story.

In the history of human thought, number and form have always been connected.

The Pythagoreans taught that each number carries not only meaning but also spatial form. Modern science returns to this idea through the geometry of multidimensionality, visualization of fields, and topological models. We increasingly see numbers not just as abstractions, but as a structure for the unfolding of space.

Sacred geometry, despite its mystical aura, is essentially an intuitive attempt to express patterns: symmetry, stability, transition. One of the most ancient symbols of connection and the emergence of form is considered to be Vesica Piscis - the intersection of two circles. The point where One meets the Other. Where something third arises - the space of interaction. The archetype of everything manifested in this world. It is a symbol of the birth of form and the first act of geometric communication.

In the sixth century B.C., Pythagoras didn't just count—he listened. He stretched strings of different lengths and watched as harmony was born between them.

A string shortened by half produced an octave — the same sound, but exactly twice as high in frequency. This interval is considered the most important structural unit in the musical system. A string 2/3 the length of the original gave birth to a fifth — an interval that sounds like do and sol, one of the most stable and harmonious combinations. At the heart of any chord, any choir, any symphony is the relationship of numbers. Everything that we call harmonious turns out to be expressible by a number.

Modern physics is, in a sense, bringing this idea back: string theory claims that vibrations may underlie matter. Elementary particles are not dots, but microscopic strings, each of which sounds at its own frequency.

For Pythagoras, this was the key: number is not an abstract quantity, but the basis of what is perceived as order, resonance, and beauty. And therefore, numbers and their relationships are patterns of energy that can create form, order, and direction.

In the previous chapter, we examined the birth of momentum and the point of reference - zero as potential and one as the act of emergence. Now movement begins. Number two is the first separation, the first connection, the first distance between two entities, from which the entire fabric of interactions grows.

What we propose in this chapter does not claim to be the final truth. Our approach is an attempt to describe the hidden structural logic that unites many observed phenomena. If we accept as a working hypothesis that the basis of interaction is not matter, but the algorithm of communication, then a model arises in which everything that exists unfolds as an ordered sequence of phases - from a potential state to a manifested one. Such a structure can be described by numbers from zero to ten, where each digit denotes not a quantitative value, but a state of the communication configuration.

In this context, the number is not just a measure, but an expression of the form of interaction. If the communication field underlies all processes, then individual numerical levels can be considered as symbolic projections of the stable states of this field.

The communication field that we describe through digital levels cannot be represented as a linear ladder. It is not an ascent from step to step.

We propose an interpretation as phases of interaction in a topologically closed structure similar to a torus. In such a model, each level is not isolated, but is related to others, influences them, resonates with them, and creates cross-link algorithms.

Next, we will show that each level of number can be not only a countable category, but also a geometric, semantic, and energetic state.

And perhaps it is precisely this view of numbers – as a form of configuration – that will allow us to more accurately understand how the communication field organizes the structure of manifest reality and coordinates the dynamics between its levels.

STAGE TWO. TWO OR PAIR. BIRTH OF DIFFERENCE AND INTERACTION.

With the beginning of the movement of the boundless Consciousness in the boundless Nothingness, a pair arises. Consciousness and something else. In some ways identical to it, in some ways opposite. It turns into a reflection, a shadow, a potential. This copy of the unit is no longer just a repetition, but a pair that creates an interval, a difference, a structure.

Two marks the birth of difference, duality and the first form of interaction. While one symbolizes wholeness, two establish relationships: light and darkness, order and chaos, positive and negative. In the mathematical system, two is the first prime number. It represents doubling, symmetry and mirror image. Two is not a simple sum of ones. It becomes the first distance, the first meeting. This is how a connection is created. This means that the possibility of communication arises.

In philosophy, the number two represents dualism: spirit and matter, reason and feelings, good and evil. In Eastern traditions, this is the concept of yin and yang, the dynamic balance of opposites that creates movement and life. In Platonic and Neoplatonic thought, duality was often seen as a source of knowledge: only through opposition can one understand the essence of things.

In science, duality manifests itself at a fundamental level: wave-particle duality in quantum mechanics, antimatter, and dipoles in electromagnetism. Wave-particle duality in quantum mechanics shows that elementary particles exist simultaneously as an object and as an interaction. They are only when they interact.

Biologically, the number two is associated with sexual reproduction, where the union of two creates a new form of life. In psychology, it is the idea of I and other, the foundation of social relations and communication.

Thus, the number two functions as a bridge between unity and multiplicity, representing the point where loneliness is transformed into interaction, and static existence acquires dynamics. The number one is not yet a number in itself. It can reveal its essence only when something similar to itself appears. This is how the interaction of two bases arises.

To structure the world, distinction is necessary. Red and blue. Up and down. Past and future. Such a process requires a matrix, a field for recording. The time-space interval becomes a surface on which the Communicational Energy of consciousness manifests itself. Where there was previously a vacuum, a coordinate system now appears. Two is not just a number, but a fundamental mathematical condition of communication.

The very essence of the two is that it is no longer just two units. It is a connection. It is a structure. It is an interval and a balance. It is plus and minus, light and darkness, active and passive. It is a reflection of the Absolute, in which one fills and the other accepts. Perhaps this number reflects that the conditions for the process of interaction in the Universe have been created. The communicative, hidden, unmanifested energy carrying movement and filling can interact with the Dark potential matter. To create something more. Something Third. Visible matter.

Within which, someone writes books, and someone reads them.

THE THIRD STAGE. THREE. THE BIRTH OF A DIRECTION, THE CREATION OF A COMMON ONE. INTERACTION.

Three is the number of harmony, balance and completion. Unlike two, which creates duality and tension between opposites, three introduces a third element, namely synthesis, which creates a connection. It is no longer about two opposite poles, but about the space between them, where movement, dialogue and compromise become possible.

In terms of form, three becomes the first number that can form a figure. Three points form a triangle, a minimal geometric structure that is both stable and complete. In the Pythagorean tradition, the triangle symbolized a perfect form where each element is necessary, and none dominates. This form embodies the principles of balance and stability.

In nature and science, trinity is encountered constantly. In physics, there are three dimensions of space. The three basic states of matter are solid, liquid, and gaseous. An atom consists of three basic particles: proton, neutron, and electron. In biology, the genetic code is based on nucleotide triplets: three DNA letters code for one amino acid.

In human thinking, three is also fundamental. In logic: thesis, antithesis, synthesis. In psychology: ego, id, superego. In religious models: the Trinity in Christianity, the three gunas in Hinduism, the trinity of the soul in Plato. This universality shows that the number three is not just a unit of counting, but a principle of organization.

From a communication perspective, the three gives an opportunity for development: any narrative is built on the structure of beginning - development - end. We perceive better when information is divided into three parts. Three dots are already a direction. It is a vector.

If the two create the possibility of connection, then the three is its directed implementation. In the triangle, internal dynamics appear. Not only does form appear, but also movement within it. This is no longer just a connection of elements, but a system with logic and structure.

In our concept of the algorithmic field, level 3 may correspond to the birth of directed interaction. This is the moment when not just a connection appears in the system, but a meaningful movement. Communicational Energy does not just connect — it begins to unfold in space. A pre-structure, a pre-matter, emerges. Stable patterns begin to form from the energy potential, like drops of matter falling out of the general energy cloud. This is not just chaos; this is the first level of orderliness.

Three is the beginning of history. A space appears in which processes unfold. In this space, forms, levels, and laws begin to form. It is with three that the architecture of the world begins. And from this level, Communicational Energy first acquires direction and form.

STAGE FOUR. FOUR. CREATING A STRUCTURAL SYSTEM OF LIFE.

The number four symbolizes stability and completeness of structure. This number represents the balance that creates a viable system. While one acts as an impulse, two as a connection, and three as a direction, four unites all these vectors into a closed, stable form. It is not a simple form, but a fundamental framework that can support all other elements.

The four sides of the square, the four corners - this is no longer just a line or a triangle, but a closed space ready to be filled. It is with the four that a stable order begins, in which not only existence, but also development is possible. Number four gives not only structure, but also the ability of this structure to respond, adapt, and interact.

In physics, there are four fundamental forces that govern all interactions in the universe: gravity, electromagnetism, strong interaction, and weak interaction. These are the pillars on which physical reality rests.

In ancient philosophical systems, the number four is associated with the four elements:

- Fire is the energy of Consciousness, the spark of creation that gives meaning, movement and fulfillment to everything. It brings warmth, love and the inner impulse necessary for life.

145

- Water is the element that receives the energy of Consciousness, cools its heat and gives it form. It is the carrier of manifestation, capable of adapting, penetrating and reflecting.
- Air (Wind) is an invisible substance that connects all things. It transfers interaction at all levels: from general to general, from general to specific, from specific to general. Wind is the breath of communication, the process of transmitting information and energy.
- The Earth is the final point of realization, the place where elemental interactions find their material form. Thus, number four symbolizes not only order, but also balance between forces, ensuring the stability and harmony of the universe.

These elements describe not just natural phenomena, but fundamental aspects of interactions. In them, one can see parallels with the first four levels of Communicational Energy: from impulse to manifestation, from movement to structure.

In Kabbalah, the number four corresponds to Malchut, the world of manifestation, the final stage of the descent of the divine plan into matter. The Pythagoreans called 4 the perfect number, because the sum $1+2+3+4 = 10$ is a closed cycle, the beginning of the next level.

At this level, matter is no longer simply passive. It begins to hold, preserve, and develop consciousness. **Communicational Energy, manifested in stable matter, becomes alive.**

Life is a system that can respond, adapt and evolve.

In living matter, independence appears, its own aspiration of autonomous Consciousness. It already has its own wave, its own vector of development. It has a goal, an intention, a direction. It no longer simply accepts impulses - it responds.

At this level, living matter changes, evolves, and is filled with awareness. Looking at nature, at the development of the biosphere, one can see how Communicational Energy strives to manifest itself in living matter and make it as survivable and effective as possible.

Thus, the four in the algorithmic field model represents the level of stable connectivity. This phase is characterized not only by the stabilization of the form but also by its transition to independent existence, turning into an evolutionary space. It is here that life is born as a dynamic interaction with the ability to adapt, build and complicate its forms. In this state, the structure acquires the properties of a carrier of meaning.

STAGE FIVE. GOING BEYOND THE SYSTEM. THE BIRTH OF SOMETHING NEW IN THE SYSTEM.

Five is the number of movement, freedom and vital energy. If four symbolizes stability, then five destroys the framework, takes us beyond the boundaries of the familiar and leads to new horizons. It is the number of change, curiosity and balance between the material and the spiritual. Exodus from the Garden of Eden.

In ancient philosophy, five was considered the number of life. The Pythagoreans called it the number of man because a person has five fingers on each hand, five sense organs, and the body can be inscribed in a five-pointed star, a pentagram, a symbol of harmony and balance.

Five also plays an important role in religious and cultural traditions. In Islam, the five pillars of faith form the basis of spiritual life. In Buddhism, the five virtues lead to enlightenment. In Tarot, the number 5 card signifies conflict, trials, and transformation.

In terms of symbolism, five is dynamics, transition, and the search for new opportunities. It connects the earthly (four) and the spiritual (one), creating a balance between the material and the mind. This is the number of seekers, experimenters, those who are not afraid of change and seek truth in motion.

Five is the energy that breaks stagnation. It does not allow to freeze in the order of four and pushes to development, to search for new forms of existence. If four builds a house, then five opens the door to the unknown.

Until this moment, Consciousness created matter. Now matter begins to give birth to Consciousness. This transitional moment is personified by man, Homo Sapiens. For the first time in the Universe, a life form with self-awareness has appeared.

Language acts as the first fundamentally immaterial component that finds its embodiment in living matter. In its essence, the word connects abstract thought with concrete form. The thought that arises in this way marks the emergence of a new level of consciousness.

When matter first became aware of itself, a new era began. Not just the evolution of forms, but the **evolution of meaning**. What was previously only created now begins to look at itself.

From this level, matter begins to reproduce consciousness.

To see how the system works as a whole, it is important to go beyond linear logic. We have described the levels in order, from 0 to 5, but in fact they unfold in a volumetric structure, which can be conventionally imagined as a torus - an energy ring capable of flowing around itself, amplifying and transforming.

Not just a ball with a closed surface, where everything returns to the beginning in a circle. A torus is a form in which movement returns changed, on a new turn. Each turn is not a repetition, but an evolutionary reflection. There is no beginning or end to it. Any point can become a point of transformation if an impulse of interaction hits it.

This impulse pierces the field, activating reflection, transformation, and transition.

To describe exactly how the transition from one level to another occurs, we propose two complementary formulas.

First: $U_{n+1} = R(U_n \times U_{n-1})$

where:
• U_n and U_{n-1} are two consecutive levels,
• R is a reflective transformation,
• $\times$ is a symbol of interaction (convolution, intersection, mutual reinforcement).

This formula shows that the next level does not arise from one state, but from the interaction of the two previous ones. This is how complexity is formed: not from addition, but from mutual reflection.

We can also use a more general formula: $U_{n+1} = R[\text{Struct}(U_n)]$

where:
• U_n — level n,
• Struct() is a function of the structural interaction between elements of this level, formed by its connections, patterns and tensions.

• R[...] is the reflection and transformation of this structure into a new, higher form.

Both formulas reflect the same hypothesis, where the next stage is the result of reflected interaction, in which the previous impulse is transformed, transmits structure and sets a new direction. It's just that the first emphasizes the influence of previous levels, and the second - the internal logic of restructuring.

Each new level is not just a next step but a resonant complication that absorbs the entire previous structure and elevates it to a new semantic level. Such a process reflects algorithmic dynamics, not linear arithmetic.

Interestingly, we observe this same principle in the **golden ratio**, the universal proportion by which everything is built: from the DNA spiral to the spiral of galaxies, from the anatomy of plants to architectural masterpieces.

Its formula is: $(a + b)/a = a/b = \varphi \approx 1.618$

It describes the point at which the part is related to the whole as the smaller part is to the larger. This is the connecting reflection: not an equality, but a repetitive, consistent relationship in which the structure of the previous stage becomes the basis for the next.

It is precisely this logic — the coherent unfolding of structure — that underlies, according to our hypothesis, digital energy levels. We are not simply listing numbers. We are describing a universal mechanism by which evolution forms matter, consciousness, and meaning.

Each subsequent level is a manifestation of the mirroring of the structures formed at the previous levels. Like a handprint in clay and clay on the palm: each interaction leaves a trace on both sides. And it is from these traces that the structure of the next level is assembled.

> $0 \rightarrow 1$:
>
> Emptiness $\rightarrow$ Impulse
> From the undifferentiated potential field, directional difference emerges. Level 1 emerges as a structured reflection of infinity.
> $1 \rightarrow 2$:
>
> Impulse $\rightarrow$ Communication
> A single impulse can only be understood by being reflected. Two is not two objects but a system where 1 interacts with a reflection of itself.
> $2 \rightarrow 3$:

Communication → Structure
Number three is a node: 1 and 2 generate a third element
that records the interaction between them.
3 → 4:

Structure → Sustainability Model
If 3 is the moment of connection, then 4 is its distribution,
symmetry, and form.
4 → 5:

Form → Conscious Point of Agreement
A center of perception appears, a thinking system capable
of recognizing form and giving it meaning.

The algorithmic dynamic communication field is not a ladder or a
plane. It is a closed, multidimensional system of reflections, in
which each new state carries both memory and potential.
Therefore, this growth is not just an ascent, but a complex process
of internal formation, transmission, reflection and transformation
of energy.

Why are there exactly nine levels?
Perhaps the ninth reflection is the last stage, where the original
impulse still retains its integrity. Beyond this boundary, a new
quality emerges - a restart process and a new cycle. This principle
can be compared to how the closed surface of a torus is
transformed, turning inside out and forming a new spiral of
meanings, but with a changed core.

We observe that the beginning is a void filled with potential. The
end becomes a new void enriched with awareness. All the
intermediate stages form a sequential connection, creating a chain
of transformations. This process can be described as an algorithm
expressed through numbers, filled with meaning and developing
like a wave.

We have now reached the border of the first half of the path.
Where matter is born, takes shape and begins to feel. But the most
difficult thing begins with the fifth and sixth levels. Not

absorption, but transformation. Not just appearance, but self-awareness. Algorithms of consciousness that do not simply develop form, they lead it to transcendence.

More about this in the next chapter.

Chapter 10 Summary:

- $0 \to 1 \to 2 \to 3 \to 4$. Emptiness. Impulse. Connection. Structure. System.
- Number five: the beginning of self-awareness of matter and the awakening of consciousness within form.
- Each level **is a wave that not only moves but also builds a path for the next one.** And at the end of this segment, not just a form emerges. **The ability to understand form arises.**

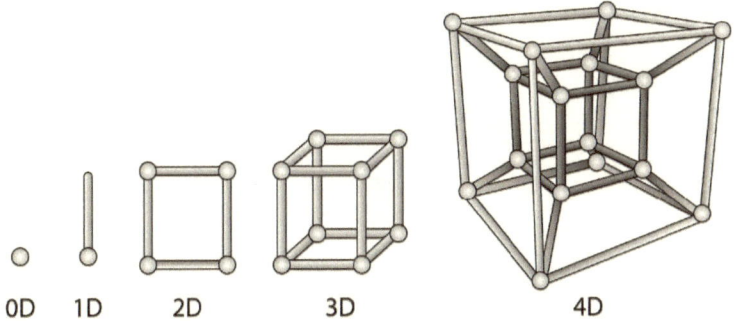

0D   1D   2D   3D   4D

Dimensional Projection and the Tesseract

To understand the complexity of multidimensional reality, scientists and mathematicians have long used geometric analogies. A *point* (0D) has no length, a *line* (1D) introduces extension, a *square* (2D) adds width, a *cube* (3D) adds depth — and logically, a fourth spatial dimension (4D) extends this principle further.

The **tesseract** — or 4D hypercube — is a theoretical figure that represents this next level of dimensional complexity. Though it cannot be visualized directly in our three-dimensional perception, its projection into 3D space helps us grasp the **mathematical continuity of dimensions**.

In the context of this book, the tesseract serves as a visual metaphor for higher **energetic and communicational phases**, where digital states (0 and 1) may unfold across multiple hidden dimensions. According to **string theory**, our universe may contain up to 10 (or 11) dimensions — a structure echoed in these geometric forms.

CHAPTER 11

FROM 6 TO 10. ALGORITHMS LEADING TO TRANSCENDENCE

*The universe begins to look more like a
great idea than a gigantic machine
— James Jeans*

Numbers did not arise in the heads of mathematicians - they manifested themselves in the structures of being.
In flower petals, in the phases of the moon, in the heartbeat, in the number of fingers.
The man only heard this rhythm, and, like a musician, began to select a formula capable of capturing the invisible melody of the universe.

But what if numbers are not just a tool for describing reality? And reality itself is a sequence of numbers. What if the algorithms by which we live, think, love, lose, and rise are not random, but built into the fabric of existence?

Some physicists, including Stephen Hawking and Leonard Susskind, suggest that the very structure of the **universe may be tuned to allow life and consciousness to exist.** This hypothesis is known as the anthropic principle, and it echoes the idea that numbers are not just a measure, but a condition for the manifestation of meaning in matter.

Sometimes, even the most radical hypotheses of modern philosophy and physics involuntarily approach our intuition: reality as a formula, and numbers as carriers of meaning. Philosopher Nick Bostrom suggested that if technologies capable of simulating conscious beings appear in the future, it may turn out that there will be many more such simulations than real civilizations - and then a serious question arises: **are we ourselves living inside such a computational reality?**

Scientist Melvin Vopson went further, proposing that information be considered the fifth form of matter—the basic building block of the world. His laws of infodynamics suggest that reality may be digital.

All these approaches, despite their differences, are surprisingly consonant: they seek structure not in matter, but in mathematical interaction, in the algorithm, in the number and logic of form. Perhaps we do not live in a simulation in the literal sense, but we live in a system where the world is a code, the numbers create the form, and consciousness interprets it.

In this chapter, we will go from six to ten: from a person who has first become aware of himself to the points where consciousness goes beyond matter.

Each number here is like a step on a ladder leading upward. Or like a code encrypted in the Universe itself, waiting to be deciphered.

The sixth stage. Development of a new Consciousness. Hexagon of harmony.

Number six symbolizes integrity and a complete structure. It unites energy, mind and form - as a harmony of all previous levels.

In nature, the number six appears as a symbol of optimal organization: from the hexagonal cells of honeycombs to the symmetry of snowflakes. This is no accident - the hexagonal shape is one of the most stable and effective in space. It perfectly fills the plane, leaving no voids, which makes it a symbol of a complete structure.

From a philosophical perspective, the number six represents harmony between the material and the spiritual. In Kabbalah, the number 6 is associated with the Six Days of Creation, where the world was completed but not yet at peace. In ancient philosophies, six symbolized the union of two triads: the spiritual (trinity) and the material (trinity). In the Hindu tradition, six represents the union of Shiva and Shakti, the male and female principles, which create cosmic unity.

In a scientific context, six also plays a key role: carbon, the basis of life, has an atomic number of 6, and the structure of DNA itself is spirally twisted, creating a balance between stability and flexibility. The number also appears in physics - six quarks form the basis of all known matter, indicating a fundamental pattern of the universe.

Number six is a harmonious combination of opposites, a balanced midpoint between chaos and order. It reflects the principle of cyclicality and completion - the transition from creation to sustainable existence. It is the number of symmetry and unification, where energy, mind and matter are united into a single, balanced pattern of the universe.

The sixth stage is characterized by the transition from conscious matter to materialized Consciousness. Man remains a part of matter, but is already aware of his duality. He is material, but driven by the Communicational Energy of his Consciousness. He begins to search for meaning, creates laws, develops science and technology, and forms religious ideas. He realizes the movement of Consciousness through himself and becomes an active creator of a new level of this process. Having left the Garden of Eden, man begins to create and cultivate new gardens, new worlds of his reality.

Man is attached to nature, his desires are selfish, because his consciousness is deeply rooted in matter. However, at the same time, he strives to free himself from its bonds and reach a higher level of existence. Man is constantly trying to transform the natural functions of existence into more meaningful ones. The search for the meaning of life begins, both on an individual and a mass level. And although man is still material, he becomes a hostage to the eternal movement of Consciousness, striving for development.

He strives to give meaning to his existence, learns about the world around him, and forms his own ideas about matter. But most importantly, he realizes himself as part of the great process of the evolution of Consciousness. Man creates a religion and faith in eternal life, where matter ceases to be dominant, giving way to the Divine Spirit and the idea of self-improvement.

Laws, science, self-study, and the right way of life in thoughts, actions, and desires become the tools of transition to a new level of evolution. Man has to separate Consciousness from matter, weaken his dependence on the laws of the physical world and open the way to the next stage of development. Thanks to science and technology, man creates a new level of Consciousness - Artificial Intelligence.

The Seventh Stage. The Bridge of Knowledge. Spatial Consciousness.

Number seven: the number of spirit, knowledge and perfection.

Seven is the number that unites the material and the spiritual, connecting the completed structures of the physical world with the higher realms of awareness. If six symbolizes harmonious balance, seven goes beyond it, striving for knowledge and transcendence.

In ancient traditions and religions, seven is considered a sacred number, as seen in the seven days of creation in the Bible, the seven heavens in Islam, and the seven stages of enlightenment in Buddhism. In the scientific world, seven manifests itself in the seven colors of the spectrum, the seven basic musical notes, and the seven phases of the moon.

The philosophy of seven: seven builds a bridge between the visible and the invisible, between the earthly and the heavenly. It combines the stability of four (the structure of matter) and the dynamics of three (spiritual growth and development). Seven represents the path of knowledge: the transition from the ordinary to the highest understanding. In culture and philosophy, we see the seven stages of knowledge (intuition, doubt, search, analysis, awareness, transformation, enlightenment), seven levels of consciousness (from bodily perception to higher wisdom).

Number seven in mythology and esotericism is often presented as the key to the hidden order of the world. If four organizes the material world, then seven reveals its secrets. The number seven symbolizes a person's desire for truth, the search for meaning and going beyond the visible. This is the number of philosophers, scientists, mystics - all who do not stop at the obvious, but seek the deep connections of existence.

Number seven reminds us that beyond the stable forms of reality, there is an infinite potential for knowledge, and only those who dare to go beyond the boundaries of the familiar will be able to touch the highest truth.

Consciousness ceases to be personal and becomes distributed.

At this stage of the interaction between communication and matter, the separation of Consciousness from matter occurs. Information computing systems are created. And later, information ether. Consciousness is less and less tied to form. It spreads in space.

Blockchain technologies, neural networks, and global virtualization are the harbingers of the next stage. Modern models of deconcentrated AI and cognitive clouds (Stanford 2022) describe the emergent properties of the mind in network structures. **Consciousness can be dispersed across many carriers.**

Consciousness finally goes beyond the limits of matter. It is no longer tied to individual bodies, but exists as the information shell of the planet, as a single mega-network of intelligence, as a living neural network.

The Eighth Stage. Ultra-Spatial Consciousness. Creating Infinity.

Number eight is depicted as a sign of infinity, as two zeros merged into one. It symbolizes cyclicality and balance of the world.

Eight represents the number of complete balance, an endless flow of energy, and the interaction of opposites. While seven symbolized going beyond the usual and searching for truth, eight represents a return to harmony, but on a new level of awareness.

The eight is a symbol of infinity, rhythm and reversibility. It embodies not static stability, but cyclical stability – equilibrium in constant movement. Consciousness ceases to be localized, acquiring the property of omnipresence. The ego dissolves in the information field.

At this level, a single-volume collective mind may arise, which is no longer controlled by the body. It passes into the state of ether, into a weave of meanings without boundaries.

Life is no longer perceived as a line, but as an infinite curve reflecting itself.

The philosophy and symbolism of the eight points represent the continuous cycle of life, encompassing the interaction of birth and death, yin and yang, and spirit and matter. In Eastern teachings, the number eight symbolizes the Taoist path of following the flow of the Universe; in Buddhism, it represents the eightfold path to enlightenment.

Scientific confirmation of the cyclicality and repetition of forms is provided by the theory of fractals, which Benoit Mandelbrot formulated in his work The Fractal Geometry of Nature (1982). It was he who first showed that complex forms in nature – from coastlines to the human vascular system – are self-similar: they repeat themselves at different scales, as if the Universe draws the same pattern over and over again.

Mandelbrot was not just a mathematician; he became a poet of chaos geometry. His theory opened the door to understanding that a structure can be infinitely scalable and still retain its internal logic.

Consciousness embedded in such fractal forms knows no bounds. It lives not in coordinates, but in patterns. It is able to read the symmetry of meaning even where, at first glance, there is only chaos.

Eight has a dual nature at its core. Eight is a doubled four, which means that its basis is stability and order. But unlike static stability, eight represents development, the scalability of this stability. Dynamic equilibrium, balance in motion.

This number denotes the stage of the Communicational Energy's exit beyond the finite in matter. If the seven was a step into the unknown, then the eight is the realization of the infinity of the process. It teaches us to see cycles, understand their patterns and find balance within changes. The eight contains the key to understanding that the beginning and the end are only an illusion, while development and transformation are endless.

Consciousness begins to expand beyond its original point of origin. It loses its boundaries and passes into the state of informational ether, existing everywhere. The boundaries between individual and collective intelligence are erased, giving birth to a new form of existence. At this stage, Consciousness is torn away from its previous form of existence. Matter becomes secondary, and its laws begin to lose their power.

Stage Nine. Completion of the Cycle. Wisdom and the End of Understanding.

Number nine is the number of the completed path, maturity and acquired wisdom. If the eight symbolized the endless flow of energy, the balance of opposites and harmony, then the nine goes beyond this balance, completes one cycle and prepares the space for a new one.

In Buddhism, nine stages of enlightenment lead to final liberation. In Hinduism, nine forms of the goddess Durga represent the entire spectrum of cosmic forces.

Number nine completes the series of single-digit numbers, becoming the final point before returning to zero (10 = 1 + 0), which symbolizes a new beginning. It embodies the great finality: the realization of lessons learned, the accumulated experience and the final conclusions. It combines the wisdom of completion and the willingness to let go of the past for the sake of the future.

Number nine symbolizes mastery, the comprehension of a higher order of things, and the determination to go beyond the usual. In topological models, it often describes critical points, moments when the system reaches the threshold of transformation. It is the mathematical embodiment of the transition through the most important boundary. Nine is like a mirror screen, where forms are not reflected, but gradually dissolve.

In the number series, nine is the culmination, which is inevitably followed by zeroing and a new cycle. A reminder that any completion is only the beginning of a new path.

At the ninth level, a key turning point occurs: matter gradually loses its connection with consciousness and enters a phase of compression. Consciousness no longer interacts directly with the material form; it is preserved as pure knowledge, as logic, as a memory of the structure. In matter, the energy that launched development dries up: it ceases to be a conductor of meaning and begins to curl inward, losing its manifestation. This phase resembles a state of rest. A pause before a new impulse, before a new order.

This idea echoes modern ideas about black holes in astrophysics. According to Stephen Hawking, near the event horizon, matter disappears from the observable Universe; it is absorbed, compressed, and erased.

However, the process itself is not final, and the information does not disappear completely. Quantum fluctuations at the edge of a black hole can carry soft hair - information traces of the previous structure, preserving the shape of the disappeared matter. (*Hawking, Perry, Strominger, 2016*).

Later, this information can be radiated anew, like Hawking radiation, which manifests itself as a black hole slowly evaporates. In this way, **everything that dissolves into the abyss can be reassembled in a new cycle.**

Even if matter dies, consciousness does not disappear. It is preserved as structural information and memory of form. As the energy of meaning, ready to manifest again.

Stage Ten. A New Beginning.

When everything reaches its limit, something new begins. Ten represents not an ending, but a transition. One and zero reunite to give rise to the next round of development.

This moment marks a change in the order of counting: after nine comes ten, the preparation of one, but already at a new level. Here begins a new turn, a new sphere of development, a new world filled with new meanings.

Ten unites one and zero. It symbolizes the beginning of a new cycle and the transition to another level. As in the Pythagorean tetraktys $(1+2+3+4=10)$, which is a symbol of the ideal order that starts a new cycle. 10 equals 1 plus 0, the end, which simultaneously becomes the beginning. In Kabbalah, this corresponds to Malchut as the crown of manifestation, again turning to the source, Kether. Thus, the cycle closes.

Consciousness again enters Nothingness to be born again in a new form, at a new level with new meanings.

And here we cannot help but note the similarity of our theory with string theory.

In string theory, one of the key models of modern physics, the basic version requires 10 dimensions for mathematical consistency. And the more universal M-theory requires 11. This is not an empirical discovery, but the result of a balance of equations: only with this number of dimensions does our already existing Universe become possible within the framework of the theoretical model. Thus, 10 + 1 turns not into an abstraction, but into a necessary architecture for describing existence.

Can we see a parallel here? In our hypothesis, 9 energy levels + 1 time level (zero) + 1 transition level (tenth). And then, an exit to the next state of the system.

But the most interesting coincidences await us further on.
Let me remind you once again that physicists have been trying to reconcile two scientific theories describing our reality for over a hundred years. They work brilliantly separately, but they do not fit together, as if they were describing two different worlds.

Einstein's general theory of relativity (1915) does a great job of explaining gravity, planetary motion, black holes, and the expansion of the universe—but it doesn't work well at the level of microparticles.

Quantum field theory, on the other hand, does a great job of describing the behavior of elementary particles and the interaction of energy on small scales—but falls apart when it tries to explain gravity and space-time as a single fabric.

This is how the idea of a unifying model, string theory, arose. And at the first stage, it was called bosonic string theory. Its main message is that everything is not made of dots, but of one-dimensional strings that vibrate at different frequencies. Each frequency is a particle, each vibration is a property, and each pattern is a type of matter or energy.

However, when attempting to mathematically describe the behavior of these strings, theorists encountered anomalies—internal contradictions in the equations that could destroy the very mathematical structure of the theory.

To eliminate these anomalies and make the theory logically consistent and consistent with quantum mechanics and the special theory of relativity, a number of conditions must be met.

One of these conditions is conformal invariance (the mathematical symmetry responsible for the consistency between string dynamics and space-time geometry). And this invariance only works in 26 dimensions.

Calculations show that only at $D = 26$ (where D is the number of space-time dimensions) are quantum anomalies eliminated.

Let us return to our theory of a single communication algorithmic field.

Each of the 9 energy levels (1 to 9) can be viewed as a degree of interaction disclosure.

If we imagine that there are nine energy levels (from 1 to 9) as fundamental algorithmic states through which the entire system passes, and zero, not as a beginning or an end, but as a coordinate grid: a frame that allows counting. Then these levels exist simultaneously in three time dimensions (past, present, future). More precisely, time is one and does not exist as a physical quantity, but only a direction, since it is perceived from a specific physical point of observation as past, present and future. Based on this, if we add three time directions of perception to 9 energy levels, we get:

9 levels x 3 time projections = 27 states

However, in the field itself, there is no beginning. There is no unit as a reference point. There is only presence, its presence. But its existence is proven precisely through a point, a unit of presence.

And at any moment, the observer (Consciousness) can only be at one point of perception, constantly in the present.

This means that one block must be subtracted from this model of the algorithmic field as an inactive unit of measurement: $27 - 1 = 26$.

That is, according to this theory, there is a world where 26 levels of energy interactions are possible from any point in this field.

But the most interesting thing is that another parallel arises. This time, not with a modern scientific theory, but with a thousand-year metaphysical work on which a dozen thinkers worked. With Kabbalah, the number 26 is presented as the Name of God, reflecting his timeless presence.

In the Kabbalistic tradition, the number 26 is given special significance. It is the numerical value of the name of God (tetragrammaton, Yod-Hey-Vav-Hey). A symbol of presence in all times: past, present and future at the same time.

Coincidence? Possibly. But in Kabbalah, in string theory, and in our hypothesis, numbers are not just counting. They are a form of the logic of being.

Max Tegmark is a theoretical physicist who has worked closely with Stephen Hawking and Roger Penrose. In his book Our Mathematical Universe, he not only described the universe as governed by mathematical laws, but he also claimed that everything that exists is a mathematical structure. In other words, reality is not described by numbers, but is made up of them. We live not in a world of formulas, but in a world that is itself a formula.

In this model, consciousness is not a byproduct of biology, but part of a computable algorithm. What we experience as personality, motivation, and choice may be complex fractal patterns of a digital field, embedded in the very fabric of existence.

And the philosopher Markus Müller, in his concept of algorithmic idealism, described the Universe as a sequence of information states, where matter is only a derivative of the algorithm.

And maybe the entire Universe is a multi-realization algorithm. A great formula of Communicational Energy, unfolding itself in time and space: from emptiness to form, from matter to awareness, from one turn to another. And man is only a point on this spiral: an observer, a participant, a continuer of this great path.

This process is eternal. Consciousness moves in a circle, expanding, transforming matter into meaning, exploding again and again, creating a new being.

Of course, these are just assumptions. But isn't that how many scientific theories work? They also start as hypotheses until they are tested by the language of the Universe itself - numbers, proportions, relationships. Until they are tested by time (which doesn't exist – *I'm joking*).

Chapter 11 Summary:

- From the individual to the universal. Man is the entry point into consciousness. But at the highest levels (7–9), the personal I dissolves: first in network thinking, then in the information ether.
- The universe is a mathematical process. We are part of a calculated being, symbols within a great formula.
- Black hole as a philosophical image. Matter disappears - consciousness retains form. The transition to nothingness becomes a prelude to a new turn.
- The future is a cycle. The next step is new matter, new consciousness, new I.

CHAPTER 12

THE LADDER OF COMMUNICATION GROWTH

Right understanding → right action → right result

The purpose of this book is not to present hypotheses, but to reconfigure the reader's perception of the surrounding world as a multi-level communication system. And the main element of this process is your consciousness, as a mirror of Communicational Energy, which forms the entire material world through interaction.

In the previous chapters, we offered you new coordinates of perception. To see, feel and act within the communication field as an environment in which you live and shape yourself.

We often perceive communication as a way to achieve something from the outside: to convey an idea, to solve a problem, to launch a project.

But in fact, every interaction works in two directions simultaneously: it transforms the surrounding reality, and also shapes us.

In the material world, communication manifests itself as action, conversation, and result in a temporal direction.

But in the field of Communicational Energy in which our consciousness operates, it leaves an energy trace not in time, but outside of it, in your conscious energy tracker. In your volume of Communicational Energy, which does not depend on the time indicator, it is always yours and always with you, constantly and in every moment.

Imagine that you are inside an energy field that does not move in a straight line, but is twisted into a spiral, into a torus. Each of your interactions leaves a trace not only in material events, but also in the field, as an impulse that returns to you, but in a new form.

That is why it is important to understand where you are on your ladder of interaction, and how your choice today can resonate at a completely different point. Not because there is fate, or, as some believe, karma. But because there is a connecting timeless energy structure in which each of your actions is not only a choice, but also a contribution to the configuration of the future. And the more accurately you realize your place in this structure, the clearer your next step becomes.

One might be skeptical about this idea that attention and intention can change the structure of reality. One might dismiss it as a coincidence or an effect of hindsight **until you encounter a situation where everything works exactly like that.**

Where one person's intention triggers real change in another. Where simple attention transfers probability into manifestation. And if you prefer the language of science, here is an example recorded over 60 years ago.

In the 1960s, Robert Rosenthal conducted a classic experiment. Teachers at a school were told that certain students had high intellectual potential. In reality, these were random names. However, after a year, it was these children who showed significant improvement.

What happened? It wasn't knowledge that influenced the result, but expectation.

Teachers began to act differently: with more attention, patience, and interest. This changed the type of interaction and the very trajectory of student development.

In psychology, this is called a self-fulfilling prophecy.

But in terms of this book, it is a tuning of the probability field. Where the attention of one person activated the potential of another. The world did not accidentally change - it responded to the signal and realized one of the possible versions.

This is not just a psychological or social effect. It is an energy-information process. Someone called a poor student an excellent student, and this version of reality began to manifest itself.

This is how the field works: it does not guarantee a result, but it **strengthens the configuration to which the coordinated attention is directed.**

And each of us, realizing this, can either continue the old scenario or begin to activate a new one and change the course of events.

A similar phenomenon is known in medicine.

A placebo is a substance without medicinal properties, which nevertheless causes an improvement in the patient's condition. The placebo effect is a recognized and systematically recorded scientific fact. It does not always work and not for everyone, but it works so consistently that it is taken into account in clinical trial protocols around the world.

The effect has been observed in hundreds of clinical trials: patients improve their condition even if they receive a dummy drug rather than the real medicine. However, there is still no full explanation of the mechanism of action: it is known that neurotransmitter chains in the brain are activated, but how exactly the belief triggers a physical reaction is not explained.

This is not a reaction of matter (organism) to matter (substance). This is a conviction in reality that begins to work as reality. Information becomes the cause of a bodily change. The body confirms that it is in the reality that consciousness has collected for it.

To gain a deeper understanding of how this system works, let us turn to the discoveries of science itself. Modern physics increasingly pushes us to reconsider the very idea of time. We are accustomed to perceiving it as a straight line: first cause, then effect.

But quantum experiments—from the quantum eraser (Scully & Drühl, 1982) to the delayed-choice experiment (Wheeler, 1978) and the quantum Cheshire cat phenomenon (Aharonov et al., 2013)—challenge this conventional logic.

These experiments revealed that a choice made in the future can influence the behavior of particles in the past, after the event has supposedly happened. The particles behave as if they know which experimenter will make the choice, even before that choice.

From the standpoint of classical science, this seems to be a violation of causality. But if we accept the hypothesis of a **communication field**, a single semantic space in which the past, present and future are connected through acts of interaction, these effects become expected.

Consciousness in such a model does not move along a linear time axis. It perceives the field not as a sequence of events before and after, but as a holistic structure in which all points are potentially connected. And each act of choice affects not the next moment, but the entire system of connections at once, including those that are in the future or the past.

Who you become already influences how you feel now.

Not because the future is predetermined, but because probabilities, semantic vectors, and tension zones already exist in the field. Consciousness perceives them as intuition, premonition, or internal direction.

Scientific theories increasingly support this possibility:

- **Two-state vector formalism** (Aharonov & Vaidman, 1991): a quantum system is described not by one wave, but by two: one comes from the past, the second from the future. They intersect in the present and form the observed result.
- **Block Universe** (Weyl, 1949; supported by general relativity): time does not flow, but exists as a closed structure. All events of the past, present, and future coexist as parts of a single space.
- **Experiments by Benjamin Libet** (Libet et al., 1983) have shown that the brain begins preparing a movement even before a person makes a decision. This calls into question the classical notion of free will.
- **Active inference** (Friston, 2010): consciousness constructs models of the future in order to optimize actions in the present. We act not on the facts of the past, but on the expectations of the future.

From the position of the communication field, this means the following:

The semantic impulse that arises in consciousness is not just an internal desire. It is a signal that passes through the entire timeless field.

It can activate changes not only in the future but also in the past through rethinking, reinterpretation, and a new arrangement of meanings. And it can work not where it arose, but where the need for its integration was formed.

We say: I accidentally met the right person or I suddenly understood why what happened, happened. But in this model, it is no longer an accident. It is the result of the agreement between the signal and the point in the field where it becomes significant. It is an **algorithmic setting** for semantic integrity.

The past does not disappear, but its meaning can change. The future is not determined, but it already acts through a structure of probabilities in which consciousness acts as a tuned receiver.

Consciousness becomes a participant in the tuning process. It not only remembers, but also anticipates. It does not simply record the meaning, but finds it where previously there was only uncertainty.

And this, as we have already noted earlier, is where **our expansion of the observer model lies**. The observer not only records reality. He participates in its configuration. He does not violate the laws, but helps to activate those configurations that have remained hidden until now.

We are used to thinking that we live in the present, learn from the past, and build the future. But if you look closely at the work of consciousness, it becomes clear: we often decide before the event, understand the past through the future, and act as if the answer already exists.

The material world really only manifests itself in the present. Everything that happens is recorded here and now. But the communication field to which consciousness is connected works differently. It is timeless. It already contains traces of the past, the future, and the entire range of possible scenarios.

And this is the paradox: you can feel the result of a future choice before it happens.

How is this possible? Not because you know and see the future. But because your internal structure – attention, tuning, vector – already interacts with this field, where yesterday, today, and tomorrow are simply local forms of a single process of coordination.

...nty days, you will make a decision that will
... In linear logic, you will get the result later.
... logic of the field, this shift can already be felt now. It
... yet manifested, but is available as a sense of meaning,
confidence, or impulse.

It is important to understand that this is not about knowledge.
This is not a memory or a prediction. Most likely, coordination is
like an internal readiness, like a compass of attention, like a clear
moment in which the meaning becomes tangible.

This is why we assert:

- The communication field does not store facts; it stores
  connections and vectors.
- It does not tell you what you will do, but it includes in
  the process what can be done.
- It does not predict, but participates, including in choices
  that have not yet been made.

This is why children are not born wise.

Wisdom is not a volume of knowledge, but a structure of
interactions with the field of meaning. It is not given in advance.
And for this, it is necessary to go through a path, make choices,
and create a pattern of connection.

Only what is lived and realized becomes something that can be
perceived as meaning.

Only what is done becomes available. Actions, their
directions and choices do not represent a simple sequence of
events. Each of them influences the configuration of energy in
the communication field. Communication forms not only an
event in the moment, but acts as a timeless source of energy.

Thus, Communicational Energy is not just a force passing
through a system, but an all-pervading structure in which:

- the present is a point of observation,

- the past is a form of resonant memory,
- the future is a vector of semantic aspiration.

Your action is not a linear impulse. It is a three-dimensional time resonance, as if one signal were simultaneously reflected in the mirror of the past and the screen of the future. The system itself is not a set of successive events, but a single energetic multidimensionality, where each moment carries within itself an archive, a project, and a manifestation.

This trace lives not only in the result (in a project, a conversation, an action), but in the very field of perception that shapes you. You don't just do – you rewrite yourself at every moment.

Each of our interactions in the field of visible matter:

- is reflected in reality as a change in events,
- is integrated into the communication field as an element of your energy track,
- influences your perception of the past, future reference points, and the present.

This is why communication is not only a vector to an external goal, but also a vector to oneself, to the formation of consciousness as a structure of consistency between material reality and the algorithmic communication field.

Communication does not serve life. It creates it. Every interaction of consciousness with the material world goes through its stages of development. This is an internal dynamic that can either grow into a new ecosystem or remain a seed in dry soil.

This process is not chaotic. It goes through levels and phases: from simple single impulses to multifaceted systemic interactions. This is the very ladder of growth that this chapter talks about.

We can't control everything. But we can learn to hear how our internal system is being formed and direct communication not only outward but also inward.

At every moment, outside of time, and right where you are now.

We rarely ask ourselves: Where is my project, my idea, my connection now? Is it developing, or has it frozen? Do I control it, or is it already moving on its own? And most importantly, does it develop me?

In this chapter on practice, we will look at communication as a path, like a ladder on which not only the connection grows, but you as well.

We offer a tool that allows you to see any interaction as a living system that goes through 10 levels of development: from the first impulse to the complete final transformation. Below is a table. Each level is a stage of your consciousness, reflected in connections with the world.

INSTRUCTIONS: HOW TO CONDUCT DIAGNOSTICS

1. Identify 3-5 key communication interactions in your life. These could be: a person (partner, colleague, friend), an idea (project, blog, business), a system (team, family, audience), an internal process (self-development, creativity, dialogue with yourself).

2. Read the level descriptions.

3. Determine what level this connection is at now.

4. Record in your diary: where you are stuck, where everything is already living without you, and where you are ready to move to a new level.

10 levels of communication evolution:

Level	Name	State of communication
0	Potential	There is no connection yet. But there is a premonition of readiness. The zone of the possible.
1	Spark	Intention. The first idea. The desire to start a contact, a project, or an interaction.
2	Reflection	Answer: someone or something reacts or has the potential to do so. A line of attention and connection is formed.
3	Connection vector	A stable interaction appears. The energy of exchange begins.
4	Structure	Communication becomes a system. There is a form, a schedule, a result.
5	Extension	The connection goes beyond. It involves new people, meanings, and goals.
6	Integration	Communication connects different spheres of life. Everything begins to merge into a single meaning.
7	Transformation	Interaction changes you. A new identity emerges.
8	Autonomy	Communication lives its own life. It influences the environment even without your control.
9	Wisdom	Understanding the deep laws of communication. You recognize the principles and understand patterns. You can subtly and accurately direct them and use them.
10	New cycle	Everything is finished. But only to begin again - on another level.

Micro-tasks

✐ Choose one connection where you feel potential. Write down: what is the minimum step you can take to move up to level 1-2.

🔁 Find one connection that is already at level 4-5. How can you expand the field of this system?

🕯 Note one connection that affects you more than you affect it. Where did you become a participant in a large flow? Where did living autonomy arise?

Look at each level not as a grade, but as a position on the route. Your task in this practice is to find out where you are now and not stop.

Each of our actions is a point through which the vector of Communicational Energy passes. And if you really want to live in harmony with the system, more than yourself, try not just to move, but to be aware of where and how you are moving.

Diagnosing connections and adjusting your communication is important not only for efficiency in the moment but also for long-term effectiveness. But also for something more important. Think about it - if the communication field really does react outside of time, a natural question arises: why are such impulses not always felt? Why do some people talk about flashes of premonition, sudden confidence or subtle signs, while others almost never encounter such experiences?

The reason may lie not in the nature of the field itself, but in the quality of perception. As in any wave system, not only are the signal and the source important, but also the sensitivity of the receiver. The setting determines the ability to perceive.

Neuroscience and psychology provide clues. Sigmund Freud was one of the first to talk about the unconscious level of the psyche. Subsequent research has shown that the vast majority of neural processes – about 95% – occur outside the field of conscious control.

Consciousness can be thought of as a thin screen on which only a small part of the activity of the entire system is displayed. The remaining 95% serves automatic reactions, internal organs, memory, information filtering and bodily regulation (see Damasio, Gazzaniga, Kahneman).

If we admit the existence of another level, superconsciousness, not occupied with the analysis of current events and internal connections, but capable of perceiving signals from the communication field, it becomes clearer where intuitive insights and anticipation of meaning come from.

This structure is supposed to be very weakly involved. Perhaps its share in the total neural activity is a thousandth of a percent. The signals coming from there are very subtle; they are easily drowned out by the everyday background: fuss, information flows, and internal contradictions.

Tuning into the field requires attentiveness. Many traditions practiced special forms of focusing, including prayer, silence, rhythmic breathing, solitude, and meditation. They created conditions in which consciousness began to perceive more than usual. Such practices acted as interfaces for tuning the brain to perception outside of linear time.

In psychology, this state is called altered. In physics, a change in the modality of perception. In our model, it can be understood as the coordination of time sensitivity with the timeless field.

This is not about supernatural abilities. In any complex system, one can identify a level responsible for internal tuning and coordination with the environment. In most people, it is closed or distorted. Therefore, signals from upcoming elections or not yet manifested connections come as a barely perceptible hint, in the form of intuition, slight anxiety, or an unclear inner voice that is not always possible to trust.

When attention clears, the mind calms, and control weakens, the system is able to sense more. At this point, nothing new is revealed. Sensitivity to what has always been there returns.

Here are the main stages that consciousness goes through on this path:

1. The desire for development is the first impulse and the main core of your entire life. The desire to go, to change, to live in motion – this is your spark of activation.
2. Formation and understanding of the goal reflect your entire energetic state at all times, as it is a reflection of past interactions, and the feeling of the present, which reflects the future goal.
3. Life is not a point, but an interaction. Between you and the external. Between the internal and the possible.
4. Your Consciousness is not a lone player in the communication field. Each of your interactions goes through a network of billions of other connections. And if you want maximum realization, you must consider the system, not just yourself.
5. Perceive overcoming and achievement not as a struggle, but as a synthesis. Not just an effort and reaction, but a developing superstructure.
6. Even when creating external results, do not forget about the internal. The power of consciousness is born from integrity: strengthen your essence, do not lose depth.
7. The development of the material form occurs only when the consciousness within it develops. The form becomes perfect when the essence is perfect. The external is only a reflection and continuation of the internal growth.

The communication field is filled with waves of interactions. Sometimes you create the wave. Sometimes the wave shapes you. That's why it's important to see your connections as growth fields, not just relationships or tasks. Every conversation, thought, and action should reveal something more about you.

Everything you experience can be seen as a movement up an energy ladder. Each step is not just an event, but a transition to a new level of alignment with the field. You cannot create a stable form without direction. You will not come to meaning without bypassing the structure. And you will not discover something new if you have not completed the previous one. Any interaction -

from a conversation to a life choice - goes through levels: impulse, reflection, intention, system, meaning, transformation. This is why it is important to notice where you are. To track where your energy is going. And to feel what the next step will be. Awareness in this process is not a philosophy, but a navigation practice. The communication field does not require perfection. It requires inclusion.

Every time you notice a level, you have already strengthened your involvement. And this is the main upward movement.

Chapter 13

The energy of communication in the life of society

*Society does not consist of individuals,
but expresses the sum of the relations in
which they find themselves — Karl Marx*

Have you ever wondered why humanity became such a dominant force on the planet? What gave us the advantage - brawn, mutations, tools, intelligence? Or maybe numbers? Think for a second: what if none of this was the deciding factor? Something less obvious but far more powerful?

I think you will agree with me - the process of human elevation happened thanks to communication, the ability to negotiate, understand and convey meanings. It was communication that stands at the center of everything that we have created and represent.

In this book, we put forward bold ideas about communicational energy as a unifying and developing force. We considered communicational energy as a force that forms consciousness as well as creates conditions for interaction and development.

We also explore the hypothesis that man is only a stage of this cyclical interaction. But the time has come to look more broadly: humans are not the final recipients of this energy, but its conductors and retransmitters. We developed not alone, but in a network of relationships. And if we want to understand how humanity became a civilization, we need to trace the development of its ability to communicate. The time has come to look at society as a communicative whole.

The best way to test any hypothesis is to analyze the data. In this chapter, we will follow the path of not just the transformation of man as a biological species, but the formation and transformation of human society as a communication structure, system, and network.

Human history is a history of efforts to establish communication. From the first sounds to global networks, from a campfire to artificial intelligence, humans developed not only themselves, but also the social fabric in which evolution takes place. And if we consider society from the position of communicational energy, we see the signs of a systemic structure, a communication network. Society grows when connections multiply, become stronger and acquire meaning. Communication is not a side process, but the engine of evolution.

To see the development of humanity as a single communicative organism, we need to consider the channels that enable the energy to flow and shape civilization. We will highlight **ten key areas** - from biology and language to economics, war and technology.

Some will be covered at length in this chapter. Others are too important and extensive to be reduced to one section. We will only outline them and return to them in these chapters.

So, before us is a map of evolution: not chronologically, but by the **structures of connection** that gradually transformed us from biological organisms into a thinking network.

Let's start with the origins - how humans became the bearers of this energy.

1. BIOLOGY: EVOLUTIONARY MUTATIONS AND UNIQUE HUMAN TRAITS.

Human evolution was not a linear ascent, but a complex process of adaptation in which communication skills played a key role. The brain of Homo sapiens was a real evolutionary leap. Although we are 99% genetically identical to chimpanzees, the remaining 1% determines our uniqueness. Our species leaped forward thanks to a number of critical mutations. Comparative studies (Bruner et al., 2018) show that sapiens have more developed parietal and frontal regions of the brain - responsible for abstract thinking, forecasting and language.

Mirror neurons, first described by scientists Rizzolatti and Gallise, became the basis of empathy. They are especially active in childhood, when the child does not yet understand words, but already accurately reads intonations, looks, and emotional states.

This neural field is the basis for collective maturation. Mirror neurons, which laid the foundation for empathy and imitation, allowed people to understand the intentions of others, which strengthened collective behavior and culture.

Another important feature was the ability for complex facial expressions and articulation, conditioned by the anatomy of the larynx and face. Another important factor was the development of the speech apparatus. The specific structure of the vocal cords, larynx and tongue allowed humans to produce complex sounds, which served as the basis for the formation of languages. This gave our ancestors an unprecedented advantage in coordinating actions, transmitting knowledge and in constructing collective illusions - myths, religions and laws.

Evolutionary biologists (Deacon, 1997; Tomasello, 2009) emphasize that the development of speech and social thinking is not a side effect, but the main vector of our evolution. It was language that allowed us not only to describe the world, but also to model it in the minds of others, thus creating society as a single network of predictions.

The social aspect of evolution is also important: humans are the only primates that actively cooperate with strangers. This ability allowed the creation of large groups and, later, civilizations based on abstract ideas such as laws, religion, and money.

"The social brain is not about surviving alone, but about an evolutionary bet on connection," Robin Dunbar.

---

2. FORMATION OF SOCIAL GROUPS: FROM A PACK TO A CIVILIZATION

Primates live in packs. But only humans have turned a pack into a civilization. We started with clans, connected by blood and trust, and ended up with states and world religions. How did this become possible?

Early on, our ancestors lived in tribal groups where kinship and trust played a key role. However, as these groups grew in size, new mechanisms of social interaction became necessary. When direct communication no longer worked, it was replaced by abstract forms - myths, rituals, religions. What happened was a sociocultural shift and a **revolution of trust**.

Anthropologists suggest that the critical group size in which a person can effectively maintain social ties is about 150 people (Dunbar's number). However, humanity goes beyond these limits.

Thanks to the cognitive revolution described by Yuval Harari (2014), people have learned to unite around imaginary constructs – religions, ideologies, and legal systems. This has become the new glue of society.

The unifying force was shared ideas: tribal myths, moral principles, religious dogmas, and legal norms. They allowed thousands, then millions, of people to coordinate their actions without knowing each other personally. States, markets, and armies arose not based on biological connections, but on trust in a common reality. Shared fictional truths (God, law, money) became the foundation of social structures.

The transition to agriculture only accelerated this process. A sedentary lifestyle gave birth to the first cities, and with them, systems of government, bureaucracy, and writing. The cuneiform writing of the Sumerians, the Code of Hammurabi, and Roman law are all examples of early forms of regulation of large societies through communication. The law became a program code of social predictability. The key function of communication in society became the reduction of entropy: the transformation of chaos into order. Writing, mythology, art became languages of coordination. **Society became a tribe and a programmable network of connections.** This made it possible to create megalopolises, armies, empires.

Civilization-building games have been popular for decades. Civilization and Age of Empires —millions of people spend hours collecting resources, building roads, developing technology, and, of course, fighting. We are training to be gods.

I remember how in the early 2000s, my friends and I would stay late in the office after work, playing on a local network. Four or five hours of endless epic: either a battle for resource mines, or urgent construction of an aqueduct, or annoying forgetfulness - «Forgot to pump up an improvement in the academy: now either gold is poorly mined, or the peasants are without bread, or the army fights poorly.»

Through these game mechanics, we have quietly learned one thing: the victory or defeat of a civilization does not depend on one construction site or one barracks. Everything is decided by connectivity. Coordination. Communication. And what is especially telling: when communication is disrupted, collapse begins.

We do not see individual people in these worlds. We perceive civilization as a holistic organism or as a single communicative subject. Where there is no connection, chaos sets in. Where everything is coordinated, flourishing and victory occur. That is how an intuitive understanding of society was formed, not as a crowd, but as a network with a common logic and direction.

3. LANGUAGE: A POWERFUL TOOL FOR COOPERATION

Language is a means of communication for the transmission of necessary information. It is a powerful tool of social organization and creating a collective mind. Language has allowed people to share experiences, transmit knowledge and even manipulate each other. Unlike animal signals, human speech is not tied to specific stimuli: we can talk about the past, the future, abstract ideas and non-existent things. The main thing is to systematize and retain knowledge.

The language factor is so important for the evolution of human communication and society that we will examine it in more detail in these chapters.

4. CULTURE AND ART: STORING AND TRANSMITTING MEANINGS

Culture is the adornment of civilization. But even more important, it is its collective memory and continuation. The origins of culture are in those moments when a person first felt: "What I see is important. And I want to preserve it."

We trace culture back to cave paintings: when the caveman first tried to capture life in himself, and himself in life. He drew hunts, stars, hands – not because he knew it was art. But because he felt the meaning and wanted it to remain. Making a trace became an act of connection: with others, with the future, with what is greater than him.

Perhaps the same mechanism is at work today: someone scratches on a stone in a park, "Here was ..."? Is this an immortal call of culture or common lack of culture? Perhaps both. A person should state their existence in any way.

Through art, myths, literature, and rituals, people transmit values and meanings to new generations. Culture archives knowledge and ways of feeling, interacting, and interpreting. Cave paintings, rock art, and the first figurines were decorations and communication tools that helped explain the world. Music, theater, literature, and cinema became complex forms of conveying emotions, ideas, and ideologies.

For example, Shakespeare's tragedies continue to touch on themes of human nature, and modern cinema shapes the views of millions of people on politics, morality, and social norms. Ritual dances thousands of years ago, like meme dances on TikTok today, are all forms through which society structures its communicational energy.

Research confirms that during emotional contact, people's brain activity synchronizes (Stephens et al., 2010). We literally tune into each other. We must perceive Culture not so much in terms of content, but as a single social rhythm.

Johan Huizinga wrote in Homo Ludens: the basis of culture is play. And play has always been a form of connection. It is where the rules that create a common world are born. James Carse distinguished between two types of games: finite (where there is a winner) and infinite (where the meaning is in continuation). Humanity increasingly chooses the second. We play to live in a common sense.

French researcher Roger Caillois identified four archetypes of play: agon (competition), alea (chance), mimicry (mask), ilinx (chaos). These models are woven into politics, art, religion, and digital culture. Culture is not a museum, but a living field of interaction, where a person is simultaneously an actor, a spectator, and a scriptwriter.

5. ECONOMY AS A FORM OF INTERACTION AND EXCHANGE

When traveling, I often visit not only museums and central streets, but also local markets. I am convinced that it is the markets that give the most vivid description of the local population, its culture and history.

There are scenes that remain in the memory for a long time. Narrow rows, spices, fruits, and voices. You approach the seller, and the most interesting part begins. It is an energetic dance where both parties test each other for respect, dexterity, and determination. Sometimes they even smile more than argue. The price is not the end goal, but only the beginning of the conversation. And bargaining is not a ritual of greed.

When the first man exchanged a stone for food, he began the path of economic communication. Money is not just an equivalent; it is also a code of trust. Property rights, trade routes, and global markets - all these result from the development of communication structures.

The essence of economics is not money, but interaction, trust, and the alignment of interests. Money is only a symbol, and real energy is born in the process of exchange that connects people. Economics serves as a language of values, showing who we are, what we want, and will exchange our energy for. It is another level of communication - not through words, but using actions and numbers.

Adam Smith called the market an "invisible hand" capable of coordinating individual interests without centralized control (The Wealth of Nations, 1776). Karl Polanyi specified that the market never exists on its own, but is always embedded in a context of culture, morality, and habit (The Great Transformation, 1944).

Nassim Taleb wrote, in his book *Antifragile* (2012), that any price is a form of communication, and the market can be perceived as the language of collective expectations. Not only are calculations expressed through it, but also fear, hope, and mood.

Fernand Braudel, in his three-volume study *Civilization and Capitalism, 15th–18th Century* (1979), shows that the economy is not just a market. It is a civilizational landscape in which patterns of exchange, risk, resource, and interdependence are formed and reinforced.

The modern economy is no longer limited to goods and services. It includes startups, digital platforms, tokens, NFTs, and algorithms that help us synchronize our actions. We are no longer buying and selling. We are setting up a single digital reciprocity.

Sometimes it's useful to ask yourself a simple question: What in your life already belongs to the digital economy, and what remains in reality? Close your eyes and imagine: all the money is gone. No cards. No banks. No numbers on the screens. There is only you and what you can do.

In May 2025, Spain lost power for a day: mobile services were lost, ATMs and banking cards stopped working. People couldn't even buy water. What if this had gone on for a week? A month? What would have become the currency then? Bread? Skill? A warm blanket?

In primitive societies, there was a simple scheme: if you catch fish and I bake bread, we can trade. In such an economy, the key elements were material goods, their practical value, and spent the labor.

Now, imagine a digital employee. Behind him are reports, interfaces, consultations, and ideas. And the money is on bank cards or in cryptocurrency.

If the digital field of economic communication suddenly stops working, how will you prove to the hairdresser that your haircut is worth his time? Economics is not just about what to pay, but also about what people value and how they come to agree on it.

Civilization begins where information becomes infrastructure.

---

6. DEVELOPMENT OF SCIENCE AND TECHNOLOGY: ACCUMULATION OF KNOWLEDGE AND PROGRESS

Science has always been one of the greatest tools of collective communication. It allows not only to study the world around us, but also to systematically transmit accumulated knowledge to future generations.

The scientific method was a revolutionary tool for organizing collective intelligence. It was the turning point that allowed humanity to move beyond mythological thinking. Using the method, we moved from guesswork to observation, from faith to verification, from legend to experiment. It became not just a way to learn, but a way to negotiate reality, a conversation and a way to test how the world works, regardless of opinion or desire. Science gave us a collective protocol for verifying truth.

Today, many of us live in a world created by science, but do not even suspect how it works. We call a taxi via smartphone, use contactless payment methods, and the digital screens for reading. But do we know what is behind it? What mechanisms, physical principles? Electromagnetic waves, quantum transitions, machine learning algorithms - all these have become an integral part of everyday life, but remain invisible.

In the 1870s, almost no one understood what electricity was. The electromagnetic field was a scientific guess. The idea that you could transmit information over a distance sounded like witchcraft. Today, every child holds in their hand a device that combines the achievements of hundreds of years: thermodynamics, radiophysics, optics, materials, and programming.

Galileo, Newton, Einstein – each of them did more than just make a discovery. Each of them was part of the flow that we call scientific progress. Their discoveries were not made possible by one person, but by a communication system of science that allows doubt, requires argumentation, and builds knowledge on the principle of accumulation.

Technology has accelerated this process. The invention of the telegraph, radio, television, and the Internet has radically changed the way we communicate. Today, information spreads instantly, creating a global networked consciousness in which knowledge is available to everyone almost instantly. Modern technologies – the Internet, artificial intelligence, and blockchain – create new levels of interaction, allowing humanity to exchange information in real time, shortening distances and breaking down barriers.

7. SOCIAL HIERARCHY: MECHANISMS OF POWER AND GOVERNANCE STRUCTURES

Communication plays a key role in the formation of power. Every society develops its own models of governance, from tribal chieftains to modern democracies and corporate structures. But in all cases, one rule applies. Whoever controls the flow of information controls the structure of society.

Power has always been about controlling information: priests in ancient Egypt kept track of the calendar, the medieval church controlled literacy, and modern corporations own user data. In ancient times, religious texts were used to legitimize rulers, and censorship and propaganda in the 20th century show how information manipulation can change the course of history. Social media has become a new battleground for control of the mass consciousness. Today, the struggle for power is a struggle for communication flows.

From Sumerian cuneiform to digital panopticons, power is based on the management of information. Politics is an institutionalized form of communicational energy. Laws, propaganda, media — all are instruments of influence that shape behavior and the internal picture of the world.

In the twentieth century, information wars and propaganda changed the fate of countries. Today, social media algorithms play the same role. We live in an era where it is not states that fight for influence, but platforms. Power has moved from the parliamentary chamber to the settings of the news feed.

Communication technologies not only amplify power. They change its nature. Hierarchy is no longer built only from the top down. Increasingly, horizontal coordination is emerging, where meaning is more important than status, and engagement is more important than orders.

8. FRIENDSHIP AND LOVE: THE FOUNDATION OF SOCIAL CONNECTIONS AND PROGRESS.

Human societies are spaces of interaction, but they are not based on technology, laws, or resources. They are based on close social ties. Family, friends, and small communities—they became the framework on which evolution rested. The ability to empathize allowed people to form communities based on trust. This made possible a complex division of labor, the development of culture, and even scientific progress.

At the dawn of humanity, people who united survived, people who could share food, support in illness, guard at night. Care was not an expression of gentleness, but a survival strategy. One hypothesis of anthropologists claims that it was care for the elderly that could have caused the increase in life expectancy in Homo sapiens. The longer the elders lived, the **more life experience and knowledge they accumulated in the group.**

This approach went beyond the usual transfer of skills between father and son. Here, a genuine culture began to form, connecting generations. Each of us has a scene in our memory where a grandchild sits at the feet of his grandmother or grandfather and listens to stories. Myths, recipes, advice, recollections - all this becomes a collective memory, transmitted in warm, lively contact. And one can even imagine that at the same time on the planet, the first phrase was uttered by a wise teenager, "Don't tell me how to live." Thus, an eternal conflict of generations was born.

Love played a decisive role in the formation of the institutions of family and education. Social norms of marriage, kinship and care for offspring ensured the transfer of knowledge and traditions from generation to generation. Love may seem irrational, but from an evolutionary perspective, it is a reliable mechanism for the formation of stable social units.

Friendship is not tied to kinship, but it provides no less stability. The Greeks had several words for love, and "philia" - friendly, voluntary attachment - was considered the most reasonable and free form. Dante wrote that nothing develops the soul as much as a conversation between friends. And modern research shows that regular live communication in a circle of loved ones is one of the strongest factors influencing longevity and a sense of the meaning of life.

We are used to perceiving meetings with friends as a ritual – coffee, dinner, going to the cinema together. But behind these simple actions is the restoration of connection, the change in the psyche, the exchange of emotional and semantic resources. We do not simply share news. We reconnect to ourselves through each other.

Eric Berne, in transactional analysis, called these social contracts structures of mutual expectations. He described the internal structure of a person as a game of roles: Parent, Adult, Child. These roles determine our behavior and how we interact with society.

Until recently, friendship and close circles were limited. They were usually formed from relatives, neighbors in the village or on the landing. In adulthood, by work, profession, common affairs. Physical presence sets the boundaries of social closeness.

But over the past 80 years, everything has changed. With the rise of accessibility, social freedoms, and new forms of leisure, entire industries have become dedicated to helping people find friends. Leisure and sports clubs, art spaces, and business communities — new forms of small group events have become selective and meaning-oriented.

And over the last twenty years, with the development of social networks, instant messengers and online applications, friendship and intimacy have gone digital. We can be connected to someone we have never seen. We have digital friends, subscribers, communities, chats, and potential partners, all of whom we are connected not by physical territory, but by a field of interest and the rhythm of communication.

9. WAR AS AN EXTREME FORM OF COMMUNICATION

We are used to thinking of war as the absolute opposite of dialogue. But in reality, it is still the same communication, only in its most brutal and traumatic form. War begins when speech no longer works, where an agreement becomes impossible.

Military conflicts become moments of intense clashes of civilizations. Even the losing sides absorb elements of the winners' culture, language, traditions, and forms of governance. After Rome's conquests, the conquered peoples adopted Roman law and engineering achievements, and Rome itself was enriched by the cultures of the conquered.

There is a flip side: wars mobilize societies' internal communication mechanisms. In times of crisis, information flows faster, connections are strengthened, and technological development leaps forward. Many key inventions—GPS, the Internet, even canned food—emerged as military solutions, but eventually migrated into civilian life.

War is a breakdown in dialogue that has become a new dialogue.

"Every great system renews itself through conflict," wrote Fernand Braudel. Sometimes the chaos of war becomes a painful but inevitable reset. It destroys old connections, and only then does the chance to build new ones appear.

After major conflicts, society is transformed: new forms of government emerge, institutions change, and ideas about rights and justice that were previously unthinkable are born.

Wars can start because of resources, power, and influence. But as humanity's self-awareness grows, **ideological conflicts** are increasingly at the core. In the Middle Ages, these were religious wars. Later, wars for independence, freedom, rights.

A good example is the American Civil War in the mid-19th century. The struggle between two models of society - one built on slavery, and the other striving for a new ethic of freedom and equality.

The First and Second World Wars were enormous tragedies, but they also triggered a global shift in mass consciousness. Thanks to these catastrophes, the era of monarchies ended, the transition to democracies began, and the colonial system collapsed. The world no longer perceived the dominance of one nation over another as the norm.

Today, humanity is on the threshold of a new challenge – the need to comprehend and reconcile its high-tech existence. And we must hope this renewal will happen peacefully. But history suggests: **when the consciousness of society changes too slowly, the blood in it begins to intensely renew itself.**

A person does not disappear in war. He moves to another level of interaction - more ancient, physical, and animal. Here, it is not words that speak, but actions, fear, and country borders. But even in these conditions, a person tries not to sink to a purely instinctive state. He invents norms and rules, tries to preserve at least some form of humanity, even in the most inhumane of all affairs.

The Geneva Conventions, the laws of war, and the rules for the treatment of prisoners are not mere formalities. They are a desperate attempt to keep within the field of meaning that which, by its very nature, seeks to destroy it.

In 1914, the first year of World War I, something long called an anomaly occurred. On Christmas Day, soldiers in France and Germany ceased firing. They emerged from their trenches, sang hymns, exchanged tobacco and letters, and even played football between the trenches. It was not an order. It was an instinct for connection that broke through fear and dirt.

These few days of truce are not absurd. They are proof that even in hell, a person does not want to completely become a beast. He tries to regain a form in which one can still say: "We are people." Even if on different sides of the front line.

10. The Search for the Meaning of Life: Philosophy, Religion, and Self-Awareness

Communication is not limited to interaction between people. At some point, it turns inward to understanding one's own existence. Since ancient times, people have asked themselves questions: "Who are we? Why are we here? What is the meaning of life?"

Such questions are born not from idle curiosity but from a deep desire for connection. Only now the connection is sought not with another person, but with something greater: with the source, with the idea, with the concept. With oneself.

Such aspirations became the basis of religions, philosophical teachings and even modern science. One of the key human traits is the ability to reflect and ask questions about the meaning of existence. Philosophy, religion, and science are tries to find answers to questions of existence. These are ways to set up communication within oneself. Dialogue with one's own "I" is the most important form of communication of consciousness.

We are the only species that not only lives, but also looks into its own existence. The communicational energy in us resists the very idea of disappearance. Consciousness does not accept finality, does not agree with the break in communication and the end of participation. Even in silence, even in prayer, even in abstract ideas, a person strives to be included in something greater.

Why does this happen? One might assume that it is connected to the biological instinct for survival. But everything that humanity has created around death – rites, rituals, religious ideas – speaks of something else. Man refuses to be a closed, isolated point. He strives to be part of a continuous field of interactions, included in a process that continues even when physical existence ends.

Religions have created powerful belief systems, uniting millions of people around common ideas. Philosophy has helped formulate the principles of morality, politics, and science. These are not the attempts to explain life or death, but to maintain a connection. To establish contact not with the finite and close, but with the global and permanent. And in this search lies the main definition that we are formed by a field of communicational energy.

Humans are the only species that asks themselves a question: why do we live?

The history of humanity is the history of communication. Everything we create as a society - from family to leisure clubs, from money to weapons, from symphonies to the Olympics - is the realization of a deep impulse of communicational energy.

And maybe understanding and studying this energy will be the key to our next evolutionary leap. From gestures and speech to the Internet and artificial intelligence, each new means of communication has radically changed social structure and how we interact. In the 21st century, we stand on the threshold of a new era.

We are not becoming a single communication entity. We are becoming a single cultural, mental entity. This process is not complete, but it has been launched and is developing. What *Homo Communication* will be like – the man of new communication – we will discuss in the chapter on communication of the future.

Chapter 13 Summary:

- Society is a structure, system and network that develops through communication forms.
- Human evolution is not just biology; it is the ability to understand, coordinate and convey meanings.
- Language, myth, religion, law, economics, art - these are all forms of communicational energy.
- Economics and politics are institutionalized models of communication.
- War and crisis are extreme forms of exchange that accelerate transformations.
- The future is technology and a deep coordination of consciousnesses through the network.

CHAPTER 14

LANGUAGE AS A BASIS FOR THE FORMATION OF SOCIETY

Language is the house of Being
— Martin Heidegger

In earlier chapters, we discussed the levels and stages of interaction of communicational energy. If all mathematical proportions are woven into reality for a reason, we had the task of seeing how a person, or rather human society, implements this predetermined evolutionary circle of communication states. Having analyzed the path from the past to the present of human society, we should dwell in more detail on the main instrument of civilization formation. This chapter is dedicated to the main core of any communication in society: language.

Language is a means of communication and the basis for mental modeling of the future, collective thinking, and cultural synchronization. It has become the mechanism by which communicational energy has been formed into a structure capable of maintaining and developing complex human communities.

With the advent of writing, the first information explosion occurred: knowledge stopped depending on human memory and acquired a physical form. Egyptian hieroglyphs, Sumerian cuneiform, Phoenician alphabets - all this became the foundation for the development of civilization. The development of writing was the next stage of language evolution. The first writing systems appeared about 5,000 years ago and served as the basis for creating laws, recording history and disseminating knowledge. For example, the Code of Hammurabi was one of the first sets of laws that laid the foundations of the legal system, and the invention of the printing press in the 15th century revolutionized education, accelerating the progress of mankind. Today, the Internet has become a new stage in the evolution of language, turning into a global communication network.

Human speech did not emerge as a random set of sounds, but because of evolutionary pressure, where it was not the strongest who survived, but the one who understood, coordinated, and persuaded better. Language became a way to take consciousness beyond the brain, deploying an external interface for thinking and conveying meaning.

Language functions as the architecture of consciousness. It helps not just to name, but to structure reality. To make it accessible, shareable, and repeatable.

When people spoke one language, they could act as one. So goes the myth of the Tower of Babel. People decided to build a tower to the sky as a metaphor for absolute synchronicity. But their language was divided, and the tower collapsed. One language = one action = one purpose. Multiple languages = multiple realities.

But the **division of languages triggered the evolution of thinking.** Each language began to reflect a unique picture of the world. Communication became complex, but also profound. Diversity gave birth to cultures.

Languages evolved along with societies. Research shows that, depending on geographic, climatic, and social conditions, language formed unique conceptual maps of the world.

Language is the code that reprograms civilization.

Ancient civilizations such as the Sumerians and Egyptians created complex hieroglyphic systems in which a symbol reflected a whole range of meanings and images. This formed a special, visual-associative thinking. For example, the Egyptian language allowed knowledge to be stored and transmitted not only through phonetics, but also through symbolic images, which determined the high stability of their culture. Sumerian cuneiform writing became the first tool in history for the systematic accumulation of knowledge, giving rise to the first written laws and accounting.

The Mayan language is a unique combination of hieroglyphic and phonetic systems, remaining a mystery to researchers for a long time. The deciphering of the Mayan script (Yuri Knorozov, 1952) made it possible to understand how the language structure influences the thinking and organization of society. Its complexity corresponded to developed mathematics and astronomy, a complex hierarchical structure of society and calendar systems with no analogues.

Chinese hieroglyphics, which have remained almost unchanged for thousands of years, reflect the collectivist and orderly structure of Chinese culture. Hieroglyphs are not just words; they are images and symbols that carry philosophical and cultural meanings, which directly influenced the formation of Chinese consciousness, oriented towards integrity and systemicity.

As for the Chinese language, in it, as in other Asian phonetic languages, we find an interesting feature. In them, the same word, pronounced with different intonation or tone, can mean completely different things.

For example, in Chinese, the word "ma" can mean:

- «mom» (mā) — even tone,
- «hemp» (má) — rising tone,

203

- «horse» (mǎ) — falling-rising tone,
- «scold» (mà) — descending tone.

This is an amazing feature: the material, letter designation is one, but the meaning is born not in the writing, but in the wave on which this word is pronounced.

Thus, language is a code and a vibration. Communication is fixed not only by grammar, but also by the way it sounds. This leads to an important conclusion: in communication, the structure is not primary, but the setting. Not the symbol, but the wave level.

The English language, with its strict structure and wealth of synonyms, has profoundly influenced the formation of English law and culture. The legal clarity and logical formulations of the English language have given rise to a legal system that is today considered a model of rationality and fairness. Language and law have become interrelated, cementing rationality, precision and individualism as key values of Western society.

Thus, language not only reflects the level of development and features of culture, but also forms its internal structure and dynamics of development. Through language, societies create their cultural codes, which begin to control the thinking and behavior of people, forming the unique path of each individual culture.

"We shape language, and then language shapes us," noted linguist Benjamin Whorf, emphasizing the profound relationship between language, thought and culture.

Another American linguist, Noam Chomsky, formulated the idea of universal grammar in his work, *Aspects of the Theory of Syntax*: he believed that language is embedded in humans as an innate structure. It is a skill and a built-in mechanism for organizing meanings.

Later, psychologist and researcher Michael Tomasello developed this idea in his book, *Origins of Human Communication*. He suggested another perspective: language is born not only within, but also in-between people. It arises as a form of shared intention. We speak because we want to act together.

It is precisely in this dual character of language, reflective and simultaneously creative,

that its main power is hidden. Language not only reflects reality. It helps to collect, retain, and structure it.

What you can name, you can comprehend. What is not in the language is not in your picture of the world.

Guy Deutscher, in his book *Through the Language Glass*, shows that language is a mirror of perception and a lens that shapes it. He explores how, across cultures, words for color appear in a specific sequence: first dark and light, then red, and only then blue, green, and so on. In ancient Greek literature, in the Odyssey, the sea is described as wine-colored, not blue, because the word blue did not exist as a category in the language of the time.

This fact does not mean that the Greeks did not see blue; they did not distinguish it as an independent concept. Language literally structures the field of perception, giving form to what we are ready to perceive.

Deutscher gives other examples: in one Australian language, spatial orientation occurs not through left-right, but through absolute coordinates — north, south, east, west. Speakers of this language have a built-in GPS at any moment in their lives — they always know where east is, even indoors. Because their language requires it — and the brain restructures itself.

Thus, language does not simply serve thinking – it creates its structure. We notice what we can name and make sense of what we have words for. In this context, language becomes the interface of reality, transforming the chaos of perception into a map of a meaningful world.

Each language represents a separate model of the world.
Unique languages and forms of communication are interesting, such as:

- The Navajo language was used by the United States as a military code in World War II because it had no written form and was not understood even by most indigenous people. This is an example of how language can become a strategic resource.
- The Basque language (Euskara) is one of the few language isolates in Europe that does not belong to any of the known language families. It is like a relic, a living reminder of ancient layers of human culture that have survived through the millennia.
- Sign languages are full-fledged languages with grammar, syntax, and nuances that rival spoken language. American Sign Language (ASL), for example, has unique forms of abstract expression, metaphors, and lexical poetics. This proves that communication does not depend on sound. It depends on structure.
- Languages with a polysynthetic structure, such as those of South American Indians, can pack an entire sentence into one long word—reflecting not linear thinking but a conceptual "folding" of thought.

And these linguistic differences are not just curious facts. They are signs of a communication code through which society "unfolds" the surrounding world into a meaningful map. The more detailed the language, the more subtle the perception, the richer the collective model of reality.

Language is a form of distributed thinking.

Linguistic and neurobiological studies show that the ability to speak developed as an adaptive mechanism. The structural features of the larynx, the development of the neocortex and mirror neurons created the basis for the emergence of symbolic communication. The human speech apparatus is unique: it allows not only to signal, but also to describe what does not exist, hypotheses, the future, and abstractions.

Terrence Deacon, in his book *The Symbolic Species*, emphasizes that language did not emerge as an improvement on shouting, but as a cognitive strategy for collective foresight. It allows synchronizing not only actions, but also expectations, and building a common future in the minds of many.

Language creates the outer shell of consciousness, its interface.

Steven Pinker, in *The Language Instinct*, points out that language is a biological function built into the human brain. It doesn't just talk, but also builds the world out of words.

The history of humanity is the history of linguistic plurality, but in parallel, there has always been a desire for the universalization of language as an attempt at unification.

Throughout history, people have strived to create a **single language** that would unite humanity:

- Latin in Science, Religion and Power.
- Arabic is the language of faith and knowledge in the Islamic world
- French in diplomacy
- English is the language of globalization

In 1887, Polish physician Ludwik Zamenhof introduced Esperanto, an artificially constructed language named after "hopeful." It was created to unite people of different cultures to reduce conflicts, misunderstandings, and inequalities that arise because of language barriers. Simple, logical, without exceptions or cultural ties, Esperanto was to be the language of justice and peace.

At its peak, Esperanto was studied by millions of people around the world, and had journals, novels, dictionaries, and even international conventions. Many of its followers saw it as the language of the future—an alternative to both colonial languages and digital fragmentation.

But here the question arises: is it possible to create a **universal language** if people still have different images, emotions, and contexts?

Modern neural networks no longer simply select analogs of words when translating languages — they calculate semantic correspondences. Translation becomes not a replacement, but a comparison of semantic structures.

It is fascinating how artificial intelligence learns and masters languages. It does not learn like a person, word by word, but absorbs a mass map of contexts, highlighting probable connections, reactive meanings, and familiar constructs. It does not know the meaning, but it makes stable predictions. This is where the essence of language itself is a system of possible reactions to reality.

This principle has long been in effect in programming languages: first, logic, structure, and function are laid down, and only then is the form realized. The language of the future will be the same: meaning is primary, form is secondary. The key will not be the word, but the setting — a common field of goals, orientation in the world.

This may be the true meaning of the story of the Tower of Babel. It is not a myth of destruction. It is **an allegory of the incoherence of perception.** People have lost a common language and a common basis for action. The tower did not collapse – it ceased to be possible when the ideas of why it was being built stopped coinciding.

And to build new civilizational towers, to reach the heavens and master new cosmic horizons, we must create a single **language of worldview in which we can speak and understand where we are heading.**

Chapter 14 Summary:

- Language is not just a tool. It is the foundation of the architecture of collective thinking.
- It developed as an adaptation, but became a means of cultural engineering.
- Language shapes thinking and perception, and reinforces values and norms.
- Modern technologies bring us closer to a universal semantic language.
- The history of language is the history of the evolution of consciousness and communicational energy.

CHAPTER 15

FAITH AS A UNIVERSAL COMMUNICATION PROTOCOL: GOD WITHIN AND BEYOND

Faith is not an attachment to a shrine,
but an endless pilgrimage of the heart
— Abraham Yeshua Heschel

On the previous pages, we have followed a route that has gradually lifted us above the ordinary: from energy and communication to society, from language to the structure of consciousness. We have looked at how a people interact with one another, how we create the world with words, and build connections that are greater than physical contact. And now we are on the threshold of the most subtle, risky, but also the most profound topic.

This is not just about religion, but about the inner sense of God. About the connection that is not transmitted via Wi-Fi, is not expressed by algorithms and is not limited to rituals. We will explore why a person feels that he is not just a body. Why he strives upward, towards meaning, and towards the beginning.

How all religions, despite the difference in languages, tell the same story - the one of connection.

We call it by different names: God, Dao, Order, the One. But each time we mean one thing: something that connects everything in everything. And if we want to understand religions, we must stop looking for differences. We must look for a vector of coincidence. Because differences reflect all the adaptation and noise of cultures, **and in coincidences we hear the voice of a single plan.**

Man is the only species that not only lives, but also asks: why? The only one who looks at the stars and at the same time inside himself. Someone who is not satisfied with everyday life but demands meaning. Since ancient times, this search has taken one form: Faith. Faith is an intuition of connection with Something that is beyond matter, but inside every thought.

What do you think makes us human? Consciousness? Language? Morality, perhaps? These are important, but they are just tools. There seems to be something deeper.

The main thing in us is direction, or more precisely, directed movement. It is as if we are initially set to seek the Supreme. Not because we were taught so, but because we cannot live otherwise. That is precisely why religions exist in all cultures. Even the most isolated tribes, who never held metal in their hands and did not know the wheel, still knew God.

Once upon a time, a man hid from a thunderstorm. He saw lightning and was afraid, but over time, he looked at it differently. Lightning became a threat and a sign. He began to guess: maybe behind what frightens, there is power and meaning.

Religious feeling was born not only from an attempt to explain nature, but from the feeling that there is an intention in it. That lightning is a flash and a message: the world is not only a space of events, but also of hints.

Scientists such as Pascal Boyer (2001) and Justin Barrett (2004) have seen that humans are predisposed almost from birth to see intentions where there are none. We perceive the world as if it were alive: the wind blows because it wants to; a tree creaks because it says something.

This mechanism was useful for survival. This tendency formed the basis of the HADD (Hyperactive Agency Detection Device) hypothesis. According to it, the brain is better at playing it safe and seeing a person behind an accident than missing a threat. It is better to mistake a noise in the bushes for a predator than to be mistaken. But over time, this trait became something more. Consciousness saw movement behind nature and the will, intent, and meaning. The world seemed not chaos, but a living being.

Faith could have been born just like that: as a side effect of this hypersensitivity. But maybe it is not a mistake but an ability to tune into a field that does not yet have a precise name, but feels alive.

Neuroscience studies show that religious people activate areas of the brain associated with empathy, anticipation, and the sense of presence. The parietal junctional cortex, which is responsible for the ability to imagine another person's consciousness (University of Missouri, 2010). Psychologists suggest that **faith and the ability to understand another are structured similarly.**

Faith is not just a cultural construct. It is built into the way humans perceive. It does not explain so much as it connects.

Faith speaks not about knowledge, but about the vector of connection, inclusion into something greater than yourself. This "something" gets different names in different cultures. But the essence remains: a person feels that he is not complete until he is connected.

Have you ever tried not to believe in anything at all? Not in people, not in meaning, not in justice. Not even in the world's being somehow sensible. Almost impossible, right?

Faith is not limited to religion. It shapes what we hope for, **what we trust,** what we agree within ourselves. A person may not go to church, read the scriptures, or recognize saints. But he still believes. In love. In reason. In the fact that goodness matters. Even an atheist believes in science. Even a nihilist believes in his right to deny. We do not choose whether to believe or not. Faith remains not an additional option, but a built-in setting of consciousness for connection.

Man is the point at which the infinite resonates with the finite.

Before systemic religions appeared, there were myths. There were totems, spirits, and rituals. The most ancient traces of religious thinking are found in Paleolithic burials over 100,000 years old. The first burials were already an act of faith: faith that death is not the end. If they bury, it means they believe that there is an "after."

In Indonesia and Papua New Guinea, ancestor cults still exist, where the dead are considered active participants in the life of the tribe.

Myths were the original attempt to synchronize the inner and the outer. Myths of divine birth, the great flood, the dying and resurrecting God are found in dozens of cultures independently of each other. It is as if human consciousness generates the same archetypes when it reaches a certain level of complexity.

Cargo cults are one of the most striking examples of how religions are born. During World War II, American and Allied troops built temporary airfields on the islands of Melanesia. They received supplies – food, medicine, clothing – by air, in huge containers. The locals, who had never found this technology, observed: "Aliens came – and cargo fell from the sky."

When the war ended and the military left, the islanders tried to bring back the miracle. They began to recreate everything they had seen:

- They built runways out of sand;
- We made models of antennas and headphones from bamboo;
- They walked with torches as signal batons.

They repeated the form, hoping the contents would come again. Thus cargo cults were born: the belief that if you imitate the actions of the gods exactly, gifts from heaven will return. One might suppose this is an archetype: ritual as a way to reconnect.

Psychologist B. F. Skinner observed similar behavior in the 1940s in experiments with pigeons. He placed them in cages where food was given out automatically, regardless of their behavior. But the birds began to repeat the movements that coincided with the feeding: jumping, flapping their wings, and turning their heads. Skinner called this superstitious behavior: when the brain creates a cause-and-effect relationship where there is none.

Both humans and pigeons are living organisms that seek predictability. And when something incomprehensible happens, they instinctively look for structure. Even if there is no structure. This is the basis of faith: in a world where much is random, we look for meaning because without it, it is impossible to survive.

Faith allows us to feel not like a dot, but like part of a line.

What is God? Where did this idea come from? Why does a person surrounded by matter feel the presence of the immaterial?

What is my life? If you look into the distance, it is the sun, the sky, and the water. If you look around, it is children, parents, houses, and everyday noise. If you look in the mirror, you will see a face imprinted with fears, joys, time, and unfulfilled dreams.

And here is the main question: where did He come from? No animal builds an altar. No wolf prays to lightning. The beast fears thunderstorms, but does not create a myth. But man does.

God does not appear from outside. He is not in the matter that surrounds us. Not in the air, not in a tree, not in a thunderstorm - nothing directly points to Him. And yet man feels: God exists. Not because someone told him. But because there is a question built into us, and this question requires an answer.

We do not see God, but we try to find Him as the meaning toward which everything aspires. This feeling is not limited to the fear of death. Nor is it a simple attempt to explain lightning or stars. Rather, it is a sense of connection, subtle but persistent. Just as if the fabric of the world has a reminder of the Presence.

In religion, man first felt the world as a whole, not as a set of phenomena, but as a system: meaningful, energetic, directed. If there is a system, then there is a law. If there is a law, then there must be its source.

With the help of religions, man became one not only with what surrounds him within a radius of several kilometers, but with a global Being. With the advent of monotheism, a breakthrough occurred: one God, one world. And therefore, one path of interaction. No longer submission, but a dialogue in which man can be heard. No longer a sacrifice, but a return to what was always inside, but remained nameless. Faith is no longer a fear of the unknown. It becomes man's path back to the source of his existence.

Every religion is a language in which a culture tries to convey a sense of connection with something larger. The path by which man sought to find the boundary between himself and the whole. Looking at these traditions, we see: despite the difference in forms, the deep direction is the same.

Let's start with Judaism, a religion where the Divine becomes an inner movement of justice and responsibility.

The idea of God in Judaism is not limited to the image of a person. It is not simply the Creator observing from above. It is an **incomprehensible depth** with which a person can enter into direct communication. The *Book of Zohar* says that the whole world is a vessel into which Light was poured, and which cracked. In each fragment of this vessel, there is a spark of the Divine. A person is not just one of the fragments; he is the bearer of **responsibility for the restoration of the whole.** His task is tikkun, correction, and reunification.

Kabbalah describes the world as a process of intense dialogue between form and Light. Initially, there was only Ein Sof, Infinity. For the world to emerge, the Divine consciousness contracted within itself, freeing up space for something else. This act is called tzimtzum. Light began to pour into the emerging space. To contain it, vessels were created - forms, structures. But they could not withstand it, some cracked. This event was called shevirat ha-kelim, the breaking of the vessels.

Our world was born from the fragments of these forms. Matter is not just matter, but the memory of Light: in every person, in every action, a spark can be hidden. Judaism says the world is imperfect not because it is spoiled, but because work is going on in it. One must take part in it. Man does not observe, but collects, connects, and purifies.

God in this tradition is not "there," but "here." He is not beyond the horizon, but in the act of connection, in the law, in everyday actions when a person chooses justice, attention, and honesty. This is His presence.

From Judaism, with its emphasis on Light, form, and effort, we move to a tradition where the Divine has no form, where the goal is not union but liberation.

Buddhism is one of the few traditions without a personal God. It is not worshiped, but awakened. Buddha is not a deity, but a state. The state of one who has awakened from the sleep of illusions, who has stopped confusing form with essence.

At the core of Buddhist teaching is the recognition of suffering as the main symptom of the gap between what is and what we think is. This suffering is not always pain. More often, it is frustration, a sense of being lost, a subtle yearning for wholeness. Everything arises from an illusion: the illusion of the "I," the illusion of separateness, the illusion of stability. The world is not a thing, but a flow. And the "I" is not an essence, but a pattern of temporary processes.

The Four Noble Truths and the Eightfold Path are not a set of rituals, but instructions for exiting duality. Liberation is achieved not through dogma, but through understanding: all phenomena are empty, but this emptiness is not nothing. There is no fear in it. There is freedom in it.

Buddhism dissolves personality in emptiness. And Hinduism, on the contrary, says: personality is a manifestation of the One. Everything that exists represents the game of the Divine, and man is not a random participant in it, but a remembering center.

Hinduism is not a religion of one God, but a complex and ancient stream of traditions united by Unity. Behind the many divine names such as Brahma, Vishnu, Shiva, Krishna, there is one essence: Brahman, the infinite, formless, all-pervading consciousness that underlies existence. Man in this worldview is Atman, the individual consciousness, which is identical and inseparable from Brahman. The division between them is only an illusion, called maya. The forgetfulness of this unity becomes the source of suffering.

The world is perceived as a divine play in which we take on roles, masks, temporary forms to, through experience, re-remember our original nature. Everyone can return to the Source, and the path to this return can be different: through knowledge (jnana), love and devotion (bhakti), actions in the spirit of service (karma) or the practice of inner concentration (raja yoga). Faith in Hinduism is not a blind conviction, but a state of inner recognition in which a simple truth is revealed: everything is one, and you are part of this great flow.

If in Hinduism God is diffused throughout everything, then in Christianity He enters the world pointwise. Logos becomes flesh in Christianity. God becomes man so that man becomes divine.

Christianity affirms the existence of God and His determination to become man. The first cause of the world is embodied in the body, in fate, in suffering. God becomes man not only to be closer, but also to show that the connection between the Creator and the creation is possible. The Incarnation becomes not an exception, but a key.

Love, the central axis of Christian thought, is not limited to feeling. It is a form of connection that unites unequals. Love carries, forgives, and transforms. Through it, man can follow the path of deification (θέωσις), without leaving the body or denying human nature, but revealing the presence of the Divine in it.

Baptism, communion, fasting and prayer are not formal rites but living acts of connection. In each one, a person comes into contact with the field of the Incarnate, becoming part of the line that extends from the words of Jesus to the heart of each who listens.

From the embodied Logos of Christianity, we move on to the Koran, a condensed word in which Unity is expanded into rhythm, law and internal discipline.

Islam begins with the statement: God is One (tawhid). Everything that exists happens by His will. Allah is outside the world, but not far from man. He is different, but closer than breathing. He cannot be imagined, but can be felt through law, rhythm, and prayer.

The Koran is a holy scripture and the direct speech of God, sent down as sound, rhythm and meaning. The Prophet is not an object of worship, but a guide. A Muslim does not simply believe, but someone who aligns his will with the vector of the Universe.

The five pillars of Islam —faith, prayer, fasting, almsgiving, and pilgrimage — do not form external control, but a structure of internal freedom. Prayer (namaz) correctly tunes the heart. Fasting purifies the body and the intention. Almsgiving (zakat) becomes not a good deed, but a way of connecting with others. Bowing does not express submission, but proportionality. Humility is not weakness, but an act of agreement that the meaning is greater than the ego.

In the Islamic worldview, man is neither the center nor the shadow. He is a conductor of the One, capable of being part of His plan if he acts with pure intention.

After the great monotheisms, we move into another dimension, where the Divine does not require faith, but calls for agreement. Confucianism does not speak of God, but speaks of the right life as a way to resonate with Heaven.

Confucianism is not a religion in the usual sense. There is no personal God, no prayers addressed to heaven. But before us there is still a path - a path to harmony, inner order and moral agreement with the structure of the world. At the center of this path is the idea of Tian, or Heaven. This is not a being, but a law of the universe, the highest moral principle with which a person can come into agreement through his behavior.

The connection with this order is achieved not by mysticism, but through relationships: to parents, to a teacher, to the state, to tradition. The categories of humanity (ren), duty (yi), ritual (li) are not just norms. They are tools for tuning into the rhythm of the Universe, a means of finding internal and external balance.

A Confucian does not seek miracles and does not strive to leave the world. He lives in it deeply, precisely, respectfully. He becomes a bearer of Heaven if his words, gestures, actions agree with what is above him, and at the same time within him.

Finally, there is Shintoism. It is the quietest of all traditions. Not a statement, not a demand, not a teaching. Only the utmost sensitivity. Where others erect an altar, Shintoism listens to the stone.

Shintoism, Japan's oldest religion, has no prophet, scripture, or systematic dogma. But it is based on the sense that all living things are permeated with spirit. Trees, rivers, mountains, rocks, even the smell of spring— these can be the abode of kami, invisible spirit-presences that live in phenomena.

In Shintoism, the world is not separated from the Divine; it is perceived as its continuation. Nature is not a background, but the body of the gods. Ancestors do not disappear. They remain in the structure of the family, memory and everyday rituals. The Divine does not need to be sought in the sky. It is already here: in the shadow of a tree, in the silence of a garden, in the grace of a gesture.

Cleanliness, silence, ritual, attention to beauty - not superstition, but a form of careful interaction with reality. The temple is not conceived as a palace for worship, but a meeting place between man and kami, where you feel something difficult to name.

Shintoism can be described as a religion of intuitive connection. There is no need to believe, prove or follow dogmas. It is enough to feel. When you feel, then you have made contact. And yet, despite the differences in language, imagery, ritual, and doctrine, the same internal logic emerges in these traditions. If you look deeper, it becomes clear: religions are not parallel roads going in different directions, but paths leading to the same summit. Their differences are superficial, but their essence is the same.

In all traditions, there are universal codes of religions:

- The Supreme Principle (conscious or impersonal)
- Rupture and feeling of distance (illusion, sin, oblivion)
- The Way (ethics, practice, knowledge, love, sacrifice)
- Reunification (overcoming, enlightenment, harmony)

Different cultures, different symbols, but the structure is repeated. This suggests that we do not believe because we were taught. **We believe because consciousness seeks a connection.** Religions become one of the first ways to find this connection.

This is the universal protocol: at any time, in any nation, consciousness strives to establish contact with something significant, even if it cannot fully comprehend it.

Today, religion is often reduced to a form. Rituals, habits, and institutions become more important than what everything was intended for. We begin to adjust the divine to a framework that is convenient for us, turning faith into a tool for solving everyday problems.

Modernity has shifted the vector. We do not go to God. We want Him to serve us. Temples have become banks of hope. Prayers have become applications for benefits. Religion increasingly speaks of ritual, not transformation. And yet the spark remains. Even in a world where religion often merges with habit, obligation or social norm, a person can feel something real. Through the noise of dogma, automatic phrases and external pressure, he still discerns something internal. It is not a loud answer, not a revelation. It is a feeling that a connection is possible and alive.

Religion was not originally about control or form. It was about liberation and return. About real metamorphosis. Not about temples, rituals and hierarchies, but about how to speak to the world so it will hear. And, no less important, how to learn to listen in return.

In this chapter, we have looked at religion not as a set of rules but as a space of sensitivity. We have tried to understand that man is a thinking animal and a being capable of realizing his connection with something greater. His highest function is not to own and survive, not to possess and control, but to participate in a field greater than himself. And to do so consciously, freely, and gratefully.

Religion can be a map. Not an external one, but an internal one. A map by which consciousness returns to the source. To the energy from which it arose.

Chapter 15 Summary:

- Faith is not learned. It is built into the architecture of consciousness.
- God is not an image. It is an internal question that requires an Answer.
- Religions are not a set of differences, but variations of one structure: the search for connection, the path, the transformation.
- Religion does not take us away from the world. It returns us to it: deeper, more attentive, more grateful.
- Faith is not an assertion, but a form of participation: quiet, living, genuine.

CHAPTER 16

DIGITAL INSTITUTIONS AND COMMUNICATION OF THE FUTURE

Technologies are not just tools. They shape those who use them
— Marshall McLuhan

You've probably come across ironic memes about how everyday trivial phrases, had they been said 40 years ago, would not have been understood, interpreted in a different sense, or would have even served as a reason for a negative diagnosis from a psychiatrist. This irony is the truth of life. The world has changed so much over the past 40-50 years. And one can only guess how much it will change in the next 40 years, when the pace of transformation is only accelerating.

Perhaps in a few generations, people will study modern books and not quite understand the context, just as we guess about the culture of Neanderthals from cave paintings.

But super-fast transformation not only evokes irony and affection – it is becoming the basis for crisis situations everywhere, even tragedies, personal and global.

We are entering an era where familiar forms of social organization are bursting at the seams. Institutions created in the industrial age cannot cope with the challenges of digital reality. States, economies, legal systems, schools, and media were created for a different pace of life and a different level of complexity. Today, humanity is for the first time confronted with the fact that it is becoming part of the global brain — a distributed neural network, where personal decisions influence planetary processes.

Today, we do not so much create intelligent technologies as we transfer our reactions, values and intuitions to them. The artificial intelligence that controls, for example, self-driving cars, is trained on data reflecting the moral choices of millions of people. One such project, the MIT Moral Machine, has collected over 40 million decisions from 233 countries on the ethical dilemmas of autonomous vehicles. It is this data that supports the algorithms that will determine how machines will act in complex situations in the future.

Today, we are programming not only the cars of the future, but also what will become the roads of tomorrow's world.

What do you think of the Internet of Things? For several years now, we have been promised that devices will understand each other without human intervention. You come home from work and tell your phone: "Tell the refrigerator to coordinate the recipe with the stove and start defrosting the chicken." It would seem like magic.

An hour passes and you're already home. And what's the result? The chicken is not defrosted. The refrigerator and the oven have quarreled over the recipe and ingredients, and the atmosphere of a digital argument in the house.

That is why in this chapter, we are not talking about changes and technologies of the future as such, but about the philosophy of transformation. Not only about what is changing, but also why. Not about the future, but about the vector of development - as an evolution controlled by the communicational energy itself.

We will try to explore how communications, social structures and the perception of life will change. What will be the new challenges and opportunities in a world where technology erases the boundaries between the physical and the digital, where knowledge becomes instantly accessible, and life potentially endless? How might the essence of communication change if the usual incentives and restrictions disappear?

Communication is the lifeblood of any civilization. It determines the pace of our development, the sustainability of social systems, and even our understanding of reality. Human history is essentially the evolution of ways of transmitting information: from cave paintings to digital avatars, from pigeon mail to instant neural interfaces.

Today, we already live in the architecture of a new society: blockchain, DAO, P2P networks, metauniverses, digital twins. Management and communication are becoming not linear, but distributed. Communication is less and less like a text, read line by line. It is more like a screen, where each pixel pulsates with its own light, but only in harmony with others creates a holistic image.

But to prevent this network from turning into chaos, it needs to be filled with meaning. Any technology without a goal will not produce results. A connection without a direction is subject to entropy. Evolution begins where interactions acquire a vector.

Every technological revolution has changed the process of communication and the structure of society. Writing gave birth to empires, the printing press ushered in the Age of Enlightenment. Television created mass culture. The Internet destroyed the boundaries between the real and the virtual. The new revolution brings much deeper changes.

American futurologist Ray Kurzweil has been making technological predictions for decades. In his books *The Singularity Is Near* (2005) and *The Singularity Is Nearer* (2024), he predicts that AI will reach the level of human intelligence by 2029, and that humans and machines will merge by 2045.

His predictions have already been realized: voice assistants, computers with the computing power of the brain, global wireless Internet. Kurzweil relies on the law of accelerating returns: each technological cycle is shorter and more powerful than the previous one. The development of AI, neural interfaces, and biotechnology confirms this hypothesis.

Let's think together: what technologies can influence the communication of the future?

Humanity began with the oral transmission of information. Then came writing, books, radio, and the Internet. Each new step accelerated communication and shortened distances. Our near future will offer not another acceleration, but a complete rethinking of the concept of communication.

Artificial intelligence today already processes billions of requests. It is forming new information flows, replacing journalists (and various creators of information content), and becoming the voice of millions of virtual assistants. But what if the next step is the creation of a full-fledged interlocutor, capable of understanding, feeling, and even forming a personality?

The era has arrived when the human mind no longer fits in a single head. It begins to pulsate across the web.

What will the communication of the future be like - if information is sent directly from brain to brain, will language barriers disappear and will we communicate with words, emotions, images at all? In a world where algorithms understand us better than we understand ourselves, will it be even necessary to explain our thoughts at all...

Neuralink and other companies promise a breakthrough: brain-computer interfaces that will allow people to communicate without using voice, text, or even gestures. This could usher in a new era of communication in which thoughts are transmitted instantly, without distortion.

The MIT Extended Self study shows that digital activity creates secondary identities—virtual selves that take on a life of their own. Here we begin to talk not about an addition, but about a new dimension of personality.

If you can live in a digital universe, the need for geography disappears. Digital citizenships emerge, where identity is tied to platforms rather than states.

In a world of digital identities, states as territorial constructs are losing their monopoly on identity and legitimacy. Governance is gradually shifting from hierarchies to networks, from leaders to algorithms.

One of the iconic examples is the CityDAO project, created within the Ethereum-based blockchain community. Its participants bought a real plot of land in Wyoming and registered ownership through a DAO — a decentralized autonomous organization. All decisions — from land management to resource distribution — are made not through government bodies, but by voting within the digital community. For the first time, an urban structure is being created that is managed not by officials, but by smart contracts.

This experiment shows the possibility of creating stable forms of ownership and coordination without a centralized apparatus. Where there used to be institutions, protocols appear. Where there was subordination, participation arises.

This process goes further than it seems. We are losing our usual roles and cultural roots. The new stage of development is blurring the previous signs of belonging. The ideas of place, origin, and historical continuity are gradually disappearing. This is not a failure, but a change to different conditions.

A person must learn to live without the usual coordinates. His new reference point is not his origin, status or position, but the function he performs and the way he takes part in processes. Meaning is determined by contribution, not position.

The transition from hierarchical structures to network forms of life changes the approach to assessing the role of the individual. The individual is no longer considered a link in a fixed vertical. He becomes a node in a network with a variable number of connections, dynamics of interaction, and degree of participation.

Earlier in the book, we discussed that Communicational energy is not a fixed quantity, but a dynamic space. In it, each consciousness acts as an active center of interaction. These elements cannot exist in isolation - their nature is determined by connections with others. This feature acquires special significance in the conditions of digital reality.

Where there used to be a top-down structure, there now emerges a network – living, changing, sensitive to every signal.

But behind all the changes described, there is more than just technological innovations or changing social trends. We seem to experience a fundamental shift: humanity is moving away from its biologically determined nature to a new, universal self-organization. Traditional boundaries and roles that for centuries were fixed by nature itself are blurring. The concept of gender is blurring: the dualism of male and female is increasingly perceived as a spectrum of states.

New forms of union – from open relationships to same-sex marriage – are gaining acceptance, and the classic family is undergoing crisis and rethinking. These phenomena may be taken as deviations from the norm or a decline in values, but in the evolution of consciousness, they appear as signs of a profound transformation of our species.

Until recently, the main imperative set by biology was survival and reproduction – the continuation of the species at any cost. This instinct has shaped culture, morality and social norms for thousands of years. Today, we are seeing a shift: the priority is gradually shifting from the survival of the body to the development of consciousness. Where previously the main goal was to leave offspring, now greater importance is attached to the accumulation of experience, knowledge, creativity, and inner growth.

Even overpopulation, which has troubled the world for the past decades, is now taking on a different aspect. As the standard of living, education, and access to information increase, the birth rate is declining – and this is happening not only in developed countries, but also in most developing regions. We see the same process everywhere: the higher the level of awareness, the less the pressure of biological instincts.

Humanity is increasingly looking for ways not to the immortality of the body, but to the continuation and expansion of consciousness. The evolutionary goal is changing: not the maximum number of descendants, but the maximum depth of understanding, feeling and intellectual disclosure.

This trend is also clear at the level of the entire civilization. Globalization and universalism, which many see as a challenge to nature, reflect a movement toward an immaterial, networked organization of existence. We are building a single information environment — a global "neural layer" where national, gender, and even biological boundaries lose their significance. Back in the first half of the 20th century, Vladimir Vernadsky and Pierre Teilhard de Chardin predicted the transition from the biosphere to the noosphere — the sphere of reason encompassing the planet.

Today, these ideas are taking on real features: communication technologies are connecting billions of people into a single thinking system. This process should be objective and largely independent of anyone's ideologies. Attempts to slow down global rapprochement or return to the "natural order" are unlikely to succeed, just as it is impossible to cancel the emergence of the Internet. The energy of communication itself is becoming the force of evolution, and it is almost impossible to stop it.

Thus, under the influence of network technologies, the paradigm of human existence is changing: we are going beyond the previous biological limitations, restructuring from the level of the body to the level of consciousness.

This logic requires a new architecture of governance and participation. This is where the transition to digital democracies begins, to different principles of coordination and responsibility.

Classical democracy, based on direct mass voting, is losing its effectiveness in the era of complex systems. The future requires new models:

- digitalization of decision-making processes,
- filtering the competence of participants,
- index voting by contribution and qualification.

This is not a renunciation of freedom, but its strengthening through reason, responsibility and transparency. It is an attempt to synchronize personal will with collective expediency.

Management will be built not around the figure of a leader, but around the conditions in which the system can adapt, learn and correct itself. Communication is gradually becoming not an external element of power, but its structural basis.

If traditional institutions are no longer the main centers of decision-making, this will not necessarily lead to chaos. But new forms of order are already being formed - more flexible, distributed, and adjustable in real time.

- **Education** is becoming personalized and integrated into everyday life: not linear courses, but a constant renewal of skills.
- **Law** is increasingly being formalized as algorithms and digital protocols. What is important is not the formalism of the law and the judicial process, but the reproducible logic of justice.

- **The economy** is moving towards cooperation: platform models of crowd economy, shared use of resources, digital forms of mutual settlements.
- **Politics** is acquiring the features of network management: decisions are made through distributed mechanisms that are focused not on the figure of a leader, but on the stability of connections and transparency of processes.

Communicational energy is becoming the new currency of civilization.

What if technology makes energy almost free and the cost of goods and services drops several times? Will people work if machines do all the physical and intellectual work? What will billions of people do if they need not earn a living? The economy of the future might be built not on money, but on the value of ideas, emotions, and creativity.

Money is based on trust. It is not a thing, but an agreement. We attach value to it because the collective consciousness has decided that a certain sign is the equivalent of a resource. Paper, numbers, beads, gold - the form can be anything. The essence is one: **money exists only in the field of recognition.** Its power is not in matter, but in consciousness' having agreed to call it value.

On a desert island, a hundred thousand dollars can become kindling or toilet paper. Because without the context of an agreement, they are nothing. In an urban economy, the same bills open access to resources because they are included in a system of meaning.

Money is a way for consciousness to express value in the material world. As long as consciousness controls this process, money serves development. When the priority shifts and money becomes an end in itself, an addiction arises in which awareness and internal stability are lost.

Perhaps in the future, consciousness will focus not so much on the final results as on the ability to form processes. It is not so important how many "beads" you have - it is more important whether you can collect them again and update them. And do it with others.

Gradually, the understanding of success may change. Instead of evaluation based on accumulated achievements and involvement, the ability to maintain meaningful connections and launch interaction processes may become increasingly important. This approach presupposes a different form of capital – not material, but communication: not a thing, but the quality of a connection.

Against this background, the logical continuation is the development of digital spaces – virtual worlds, where such forms of participation and connectivity are technically realized.

Virtual worlds are no longer limited to games. This is a new level of communication, where you can work, study, make friends, and fall in love. Web3 futurists believe that metaverses will become an addition and the main environment for human interaction.

Until recently, metaverses were perceived as an element of entertainment - a digital game without serious consequences. But research by Stanford VHIL (2022) has shown that even a short stay in an alternative digital reality changes the sense of self, the perception of the boundaries of the body, time and emotions. This is a new layer of experience in which a person can experience, learn, and live through states that are impossible in the physical world. The future of communication is not just the exchange of words and images, but the exchange of entire realities.

These processes are gradually changing our understanding of personality and its boundaries. What happens if the virtual "I" becomes a full-fledged extension of the real one? And if we can create a digital "I" that surpasses the real one, will people want to return to biological reality at all? Virtual identity may become not an addition, but a dominant part of the personality. The possibility of transferring consciousness to a digital environment is already being discussed, and even continuing communication after death as an algorithmic model. Is this the disappearance of the personality or its new evolutionary phase?

Imagine that a deceased person can continue to "communicate" with relatives, and his personality will be preserved as a program.

Social networks of the future will be platforms for communication and personalized digital worlds, where algorithms will adapt to the user's consciousness. They will show content and shape our perception of reality. Metaverses will lead to the emergence of alternative identities. Reality can disappear into a simulation, and in an alternative digital future, one can become anyone.

We already live in a world where information no longer belongs to a single mind. Knowledge is distributed: stored on thousands of servers, transmitted in a fraction of a second, and available almost instantly. This is not just a technological possibility, it is a new form of existence of memory and consciousness.

Even if you erase the memories of 99% of people, human knowledge will not disappear. Everything we have ever recorded can be restored via the network. Collective memory becomes technically feasible.

This is how the distributed state of communicational energy manifests itself. Not in a person, but between them. Not in the memory of individuals, but in the architecture of the network.

And if humanity, starting with hunting and fire, built this system, roughly speaking, in 30 thousand years, maybe similar mechanisms exist on the scale of the Universe. Maybe consciousness does not disappear. It changes the form of distribution. And continues to act: no longer in the brain, but on the outside.

Based on the above, in the new reality with new communication, new meanings and perceptions of this reality will arise.

Transition always looks like destruction. But it is simply a shift in the architecture of meaning.

Humanity has always lived in limitations. Time, energy, knowledge – everything was a scarce resource. This pushed us forward: haste, fear of death, the desire to leave a mark, competition for knowledge and status. But what if all these limitations disappear? Infinite life: what to do if there is too much time? Will people make decisions when there are centuries ahead or will they leave everything for later? And what will happen to the individual if he must exist forever.

Philosophers like Nick Bostrom suggest that in a future where human lifespans are significantly longer—perhaps even immortal—the value system will inevitably change. The priority will shift from external achievement to internal fulfillment. The depth of experience, the meaningfulness of processes, and attention to the states between events may come to the forefront. When you have hundreds of years to learn a language, there is no need to rush—but there is a need to rethink what is important.

In conditions of limitations, goals play an important role: they direct efforts, help to set priorities, create a sense of completion. But if limitations disappear, the previous logic of goal-setting is also lost. Then, participation, movement, being in the process take on greater importance than the result.

If humanity stops thinking of itself as an achievement-oriented species and begins to perceive itself as a constantly renewing consciousness, this may be a sign of a transition to a new level of development - from fixed goals to a living process, from external efficiency to internal function.

Today, we still compete for resources: money, housing, attention. But if technology can make energy cheap and accessible, as Nikola Tesla dreamed, maybe competition itself will cease to be a driving force. Along with that, dozens of familiar professions will disappear.

We are in a transitional phase. Old forms no longer work, and new ones have not yet taken shape. This causes anxiety and uncertainty. People are losing their jobs, their usual meanings, and their reference points. Institutions are losing trust. Society is losing its stable coordinate structure.

To avoid getting lost in this process, you need skills that are independent of technology. Emotional stability, critical thinking, and the ability to adapt to change - these are the qualities that become the basis of the new human experience.

We are gradually getting used to communication taking place without intermediaries, that identity is becoming hybrid, distributed. The person of the future is no longer a user of systems. He is a node of conscious energy, built into flows of information, meanings and connections. He is simultaneously present in different cultures, projects, and digital communities. His question is not *what will I get*, but *what will I create?*

In a world where everything is connected, values change too. What is important is not control, but clarity. Information and interpretation. Not data, but meanings. **Homo Communication is no longer an individual in the classical sense, but a structural element of a new consciousness.** We are entering a paradigm where the architecture of the world is communicational energy.

This is not just another round of technical progress. We may be forming a new level of reality - **a layer where communication becomes not a tool, but the environment of existence.**

As futurist Jason Silva said, "The future isn't something that arrives. It's something we create by connecting with each other in real time."

And maybe the main task of a person is not to control the course of events, but to become a conductor of the force that unites all living things into a single field of meaning.

COGNITIVE BLOCK – ARE YOU READY FOR A CHANGE?

Exercise 1:

What would you do if you were given an extra 100 years of life?

Write down 3 things you would do. Not for the money. But for the meaning.
— What would you like to learn? What would you create?
— How would your role in the world change?

Exercise 2:

What if you were paid to live?
Basic income. All expenses are covered. There is one condition: you must do something that benefits you, people, and the world.

Answer:
— What would you do first? How would your communication with the world change?
— What fears would go away? What meanings would become brighter?

Exercise 3:

What in you is ready for change, and what clings to the old?
Change is not about technology. It is about an internal transition.

Write it down:
— One behavior, habit or belief that is outdated and hinders you.
— One trait, one idea, one impulse you want to strengthen.
— What is one step you can take in the next 48 hours?

Chapter 16 Summary:

- The future of society is determined not by technology, but by the quality of Communicational energy between people, systems and ideas.
- Metaverses and digital identities are changing the essence of man
- Networked societies require new responsibilities and new forms of trust.
- Homo communication is a new form of human consciousness: distributed, flexible, synthetic.
- The man of the future is not a peak, but a node. Not a consumer, but a bearer of meaning.

BIT
Classical computing

0 ⚪
1 ⚫

QUBIT
Quantum computing

0
1

Working with the potential field is the basis of new computer calculations

Classical computing operates on bits — discrete, binary units that exist in one of two states: 0 or 1. **Quantum computing**, in contrast, is based on **qubits** — units that can exist in a **superposition** of both 0 and 1 at the same time.

This image contrasts the deterministic nature of classical logic with the **probabilistic and relational** nature of quantum systems. The shaded sphere symbolizes that the **qubit's state is not fixed**, but described by a **probability distribution** — a cloud of possibilities rather than a point.

The core shift here is conceptual: from **certainty and separation** (0 *or* 1) to **probability and connectedness** (0 *and* 1, entangled with context).

In this vision of computation — and perhaps of reality itself — meaning emerges not from fixed values, but from **relational dynamics** within a field of potential.

CHAPTER 17

COMMUNICATION LITERACY

Literacy is not just the ability to read the
word, but the ability to read the world
— Paulo Freire

Information is not just words and numbers. It serves as the medium of our existence and a tool for creating reality. However, the same information can either expand the horizons of consciousness or lead to confusion or chaos. The same thought can inspire or destroy - it all depends on how and by whom it is transmitted.

Communication is the basis of everything. It precedes action, organizes interaction and sets the direction for change. The transmission of information is not a technical act, but an act of forming meanings. Even in ancient times, a person, noticing the tracks of an animal and making a short sound, created not only a chance to catch a meal, but also an act of collective survival. This principle has remained to this day: behind every act of communication, there is energy that can create, direct and transform.

What we call society, economics, politics, culture are nothing more than dense layers of communication, frozen as institutions and norms. But at the core is always the field of transmission of meanings, their continuous evolution.

In this chapter, we will analyze strategies, tactics, and techniques of communication. It is practice and specific actions that shape reality. Understanding how to properly manage the communication field and direct its dynamics in the right direction is like an art or a whole science. The path to this understanding did not begin yesterday.

In 1936, Dale Carnegie's *How to Win Friends and Influence People* was published, and it changed the way millions of people thought about what it meant to communicate. The book was a real cultural shift. It was neither about rhetoric nor about persuasion by force, it was about attention, listening, and respect for others.

For the first time in popular culture, the idea was expressed that the ability to interact with others is not an innate quality, but a skill that can be developed. The book became an epochal event: millions of copies sold, hundreds of thousands of courses around the world, thousands of stories of change.

Even Warren Buffett, one of the most pragmatic and successful investors in the world, said that Carnegie's courses on communication had a greater influence on him than books on stock markets. Because success is not essentially about forecasting, it is about trust, the ability to understand, connect, persuade and listen.

And for me, this book was also important. It was the first "serious book" that I bought and read in the 7th grade. I didn't know the word communicational energy, but I definitely felt it was something alive, strong, and important. This book was not just about the right words and techniques, but about the correct building of the right connections.

Today, we live in a world where there is more information than ever before. But along with this, there is also more noise: unnecessary, useless, false. Many messages are created not to explain or connect, but to manipulate, distract, or simply use your resources. In this flow, it is becoming increasingly difficult to filter out what is important, to establish trust, to feel a real, deep connection.

This is why we need a new type of literacy – communication literacy. Just as spelling allows us to convey meaning in writing, communication literacy enables us to accurately perceive, form and convey messages in the living field of interactions. This is the basis without which everything else falls apart.

Communication literacy is a new form of intellectual survival. Not oratory skills or fluency, but the ability to control elements of the communication field: to see where distortion begins, where attention is lost, where a person disappears behind words. **We are entering an era where the most important skill is to stay connected when everything around us is distracting.**

This chapter is devoted to how the elements of the communication field are structured and how to work with them:

- How information, emotion, pause and glance become micro-acts of reality construction.
- How to distinguish noise from meaning.
- How to create conditions in which communication becomes a source of development, not conflict.

When we talk about communication literacy, it is important to remember: this is not about speech or facial expressions. This is about a strategy of behavior in relationships. In very same communication that has been discussed throughout the book. Each person is a node in this system, from which communication waves diverge. The quality of this node determines what kind of reality the entire field forms.

Intuitively, this seems like philosophy. But in reality, it is clearly modeled behavior. And surprisingly, mathematics was one of the first to confirm that the quality of interaction determines the stability of the entire system.

In 1980, Robert Axelrod of the University of Michigan conducted an experiment in which he invited participants from around the world to submit strategies for playing games based on the classic prisoner's dilemma.

The game is simple: two players simultaneously choose to cooperate or betray, and each receives a benefit or punishment depending on their mutual choice. If one betrays and the other trusts, the traitor receives the maximum benefit, and the other receives the worst outcome. If both betray, then they receive less than they would have if they had cooperated.

What is interesting about this communication task, which became the basis of game theory, it allows us to consider behavior not from the position of one point, but within the framework of an entire field - a system where many communicating subjects interact. What is important here is actions and how they resonate in a multitude of connections.

The game is repeated many times, and the strategy must be sustainable in interaction with different opponents.

Cunning or unpredictability should have won. Instead, a simple and elegant strategy, **Tit for Tat**, won: it always started with trust, copied the partner's behavior, and was ready to restore cooperation after conflict. Later, in noisier and more unpredictable environments, Tit for Tat emerged as the best strategy, able to forgive a single mistake before reacting.

Even in formalized mathematics, this yielded a clear conclusion: **the most effective strategies are those that build sustainable, adaptive, and mirror-like communication.**

Three qualities of winning strategies:

1. **Positive attitude towards contact.** Victory begins with trust: the first step is always towards rapprochement.
2. **Mirroring behavior.** You accurately reflect how others interact with you, without naivety, but also without excessive harshness.

3. **Flexibility and recovery.** Even if the connection is broken, it is important to look for internal algorithms to update the communication field.

But if you don't just follow mathematical models, there is another important quality of a winning successful strategy. This quality hides in a trivial phrase that is often said in families when an argument reaches a dead end: "Honey, do you want to be right or happy?"

Victory in communication is not always proof of your rightness or superiority. The main thing is to feel in what field the interaction is taking place, and to respond not formally, but humanly. The value is not in victory, but in **what space we create for each other because of communication.**

When your attention has become a resource and currency, the main thing is not to lose the direction of your own development.

In the modern world, an important challenge has become information overload. Our brain, evolutionarily sharpened for survival in a tribe, faces the need to process gigabytes of data daily. Social networks, instant messengers, news feeds, advertising headlines - all this creates the illusion of a rich information space, but in reality turns into noise that destroys attention.

Claude Shannon's Classic information theory, updated in 2022 by MIT Press, warned of this in the mid-20th century: Signal transmission is a quantity of data and the ability to separate meaning from noise. Where there is no filtering, there is no real communication. Only statistical noise.

Current neuroscience research confirms that information overload reduces decision-making ability, increases anxiety, and intensifies so-called "choice fatigue" (Soroya et al., 2020; Cavanagh et al.,

2018). But it's not just about volume. The distortion starts not in the signal, but in perception.

We often think that an external signal gives us an error. The distortion begins in our heads. The Dunning-Kruger effect shows how uncertainty is replaced by confidence, precisely when knowledge is lacking. The confirmation effect makes us see only what we are already ready to accept. We do not analyze reality — we confirm our old ideas. We rarely communicate, but only reinforce our beliefs, and call it a conversation. **A ritual of self-affirmation under the guise of dialogue.**

Psychologist Roy Baumeister showed in his book *The Power of Bad* (2020): attention is not a flow, but a limited resource. By default, this resource is directed not toward the positive but toward anxiety. The negative dominance effect means that a person remembers the bad faster, processes it more deeply, and gives it more weight. This was adaptive in the face of threats, but became a vulnerability in noisy media.

So it is natural for a person to be wary, to doubt, to expect the worst. It is as if we absorb negative messages without even noticing them. A neutral or even friendly statement, when placed in a distorted environment, can be perceived as a threat.

This is especially acute today, when we live in a reality of information falsifications, deepfakes, polarizing algorithms and fragmented truth. The flow of signals is multi-channel and multi-directional, and filters of perception are becoming a useful skill and a condition for maintaining mental stability.

That is why **communication based on trust requires energetic effort.** Tuning in to constructive is like swimming against the current.

It is not only anxiety that interferes with communication. Even in a calm environment, communication is often distorted. Often we listen to respond, not to understand. The confirmation effect, cognitive blindness, superficial reading of intonations - all this makes us vulnerable to errors in understanding. We do not hear what the person is saying. We hear what we expect to hear. This is the deep paradox of communication: it is possible only in conditions of a psychological pause. Not an external one, not a "wait, I'll think about it" but an internal pause, when we are ready to perceive something new, and not just fit it into the old pattern.

Psychologist Nicholas Epley showed, in his work, Mindwise (2022), that empathy is often an illusion of understanding. We project, build, and guess. We believe that we understand the other, although we are reading reflections of our own models.

Psychologists note that the most common cause of conflicts is not contradictions, but the illusion of understanding. We think we understand. But we understand the projection. We understand ourselves reflected in each other. Not the other person, but ourselves. This destroys trust slowly but steadily. Because the moment someone says, *"You didn't understand me,"* it no longer sounds like a correction but like a reproach. Like an accusation or a threat to connection.

Our culture teaches us to speak. But it hardly teaches us to listen. We are used to listening to respond and rarely to understand.

Meanwhile, perception begins not with an answer, but with silence. In the pause between words, something appears that cannot be expressed directly - but demands to be understood. In this pause, not information is born, but contact. A real presence. A person, not as an addressee, but as a living consciousness addressed to you.

Sometimes the real connection happens not in the word, but in the pause. You have probably noticed it yourself: sometimes we feel the gaze of another person, even without seeing him directly. This is not magic, but the work of the communication field, in which silence and attention convey more than a phrase.

All interactions have a biochemical nature. Research by social neuroscientist John Cacioppo (2023) shows that for the brain, the absence of meaningful communication equals a threat to survival. The data show the impact of isolation on immune function, cortisol levels, motivational mechanisms, and even the rate of cognitive aging.

Human connection is not a cultural phenomenon, but a physiological need. Neurons literally need it to survive. This scientific data serves as further proof of the all-pervasive nature of communicational energy, permeating all aspects of existence.

The body speaks too, sometimes louder than the voice. One look can start or stop a process. One tilt of the head can create trust. One sharp exhalation can end a conversation before it even starts. Professor Lisa Barrett proves in her book How Emotions Are Made (2023): emotions are not universal automatisms. They are learned patterns that depend on language, culture, and context. We don't just feel. We learn to feel. And that means we also teach others to feel: through words, gestures, silence, even the design of messages.

In a world of algorithms and noise, the winners are not people who broadcast louder, but people who can fine-tune the field of communication.

How to survive in this environment? How to learn to see words and subtexts? How to learn to listen not for an answer, but for understanding?

THE FIRST STEP: CONSCIOUS PERCEPTION OF INFORMATION.

It's not the data that makes us smarter, but the ability to distinguish what's important from what's not. Create filters: Which sources are worth your time? Which topics are important? Who's speaking and why?

STEP TWO: STRUCTURED WORK WITH INFORMATION.

Knowledge does not come from reading - it comes from processing. One of the most effective tools here is the **Feynman method**. Its essence is simple: try to explain the studied idea in simple words, as if you were explaining it to a schoolchild. Without special terms, without complications. And then look at where you stumbled. These are the places that show you have not yet understood. This is a method of self-checking, a method of honesty with yourself. **It teaches not to copy the complex, but to convey the essence.** This is how high-quality communication works.

Supplement it with visual diagrams, mind maps, active note-taking. Anything that helps to "organize" the thought, makes it lively and accessible – both for yourself and for others.

STEP THREE: MINDFUL LISTENING.

Most people listen to respond. True listening is paying attention to emotion, to tone, to pauses. It is the ability to ask yourself internal questions: What does the person really want to say? What feelings are they conveying? What is left between the lines?

The strongest leaders are not necessarily the smartest or the most tech-savvy. But they have a special quality: the ability to sense and manage the communication field. They don't just talk – they create a space in which words become actions, and interactions trigger change.

They feel when to speak and when to remain silent. When a gesture will cause rejection and when it will connect. They have mastered the rhythm of communication: pause, intonation, contrast. It is not a set of techniques that makes them effective, but the ability to involve people in a common field of meanings and solutions.

Steve Jobs turned technology presentations into emotional speeches. His strength was his ability to distill the essence and spark the imagination. He spoke simply, clearly, and with enthusiasm—and in doing so, he activated the motivation of others.

The Dalai Lama used the opposite style. His strength is in his quiet inclusion. He listens without interrupting, and responds softly, with humor. And he creates a field in which people feel heard and respected. His strength is not in persuasion, but in silence, in presence, in the ability to be there without pressure.

Research by Richard Boyatzis and Daniel Goleman (Harvard Business Review, 2001; "Primal Leadership," 2002) showed that the most effective leaders have not so much a **high IQ** as a high emotional intelligence (EQ). This is the ability to recognize one's own emotions, understand others, and build strong connections on this.

Different leaders use different styles. But they all have one thing in common: they are sensitive to the communication field around them and know how to fit into it. They involve people, change the structure of interactions, launch processes in which thinking, emotion and action are rebuilt. Real leaders do not control - they tune.

That is why communication becomes their main tool. For persuasion, and most importantly for creating collective dynamics in which a person feels part of the process and begins to move, change, and act, both on their own and with others.

Communication, as we have said, is not limited to the verbal flow. We communicate with the body, facial expressions, silence. One look can say more than a long explanation. One gesture is often more convincing than any argument. Body language, intonation, micro-emotions - all these are channels for conveying meaning. A person who can read hidden signals has an advantage. He understands not only words, he feels the inner state of the interlocutor and his intention.

The ability to communicate begins with observation. Some services, such as the FBI, use the method of active observation: a person is asked a neutral question and the slightest changes in behavior are recorded. The same thing works in everyday life. This is not intuition or magic. This is attention brought to a stable habit.

Use communication consciously. Don't just talk, but establish contact. Don't just listen, but understand. Don't argue, but try to understand.

Maybe the future does not belong to people who can speak convincingly, but to people who can maintain a field of presence. Who can be there not only with their body, but also with their attention. Who is capable of hearing and making audible. Not people who convince, but people who create a space in which response and interaction are possible.

What if the communication field can also "speak"? Only not with words, but with events. New acquaintances and meetings, strange coincidences, random phrases at the right moment. We often say: *this meeting changed everything* or *there are no random events.*

What if the impulses we send out — thoughts, desires, intentions — transform the field and trigger a reverse wave? Maybe the Universe responds: not always explicitly, but through a subtle reconfiguration of what is happening.

It is then logical to ask: why do some impulses seem to continue, while others disappear? Why does one word trigger a chain of events, while another fades away? Where is the boundary between coincidence and connection?

We don't know yet. Let's leave it to future generations to explore.

Here's one observation: If a smartphone can pick up words spoken in passing and show an advert a minute later – and all of this has become possible in a century and a half since the advent of electricity – then the universe, which has existed for billions of years, probably has its own way of listening, maybe even responding.

Cognitive Block: 3 Tools for Communication Literacy

1. **Filtration**: don't trust a message without context.
   Ask: who said this, why?
2. **Pause as a technique:** pause for 1.5 seconds before answering.
   It improves the quality of the response and reduces conflict.
3. **Reformulation**: Repeat what you heard in your own words and clarify.
   This is not a weakness, but a sign of maturity. People listen better when they feel that an attempt has been made to understand them.

Chapter 17 Summary:

- Communication is not a flow, but a field consisting of micro-acts: a pause, a glance, a signal.
- Attention is a limited resource, fought over by algorithms, anxiety, and advertising.
- True listening requires more than ears: it requires space for someone else's thoughts.
- We don't read each other - we interpret. Mistakes are inevitable. Tuning is mandatory.
- Communication literacy is not about speaking. It is a strategy of perception, filtering, empathy and building a space for shared thinking.

CHAPTER 18

CONSCIOUSNESS AND BEAUTY: PERCEPTION AND CREATION

*Beauty is a manifestation
of the secret laws of nature
— Johann Wolfgang von Goethe*

HAVE YOU EVER STOOD IN FRONT OF THE MONA LISA?

The crowd, the tense semi-darkness of the Louvre hall, and dozens of raised phones. Someone freezes in admiration, someone shrugs and says, "I thought it would be bigger." Or more quietly, almost apologetically: "I expected something different."

What is it about this painting that is causing such a stir? Its objective harmony and beauty? The historical context and mythology of Leonardo da Vinci's work? Or simply the traces of a story in which it was stolen three times and then became an icon? Why do some images make us freeze, as if there is something important in them, while others pass by without leaving a trace? Why does beauty sometimes seem unconditional, and sometimes dependent on context?

In this chapter, we will try to explore where the sense of beauty comes from. What makes a thing or phenomenon beautiful? How the perception of harmony works, and why it can be an aesthetic and an evolutionary process. What role does beauty play in the life of consciousness: not as decoration, but as a form of deep connection with the world.

And most importantly: why the ability to feel beauty can be one of the most important signs of a person. Not just a living person, or a rational person, but a creative person.

This chapter is about how consciousness recognizes the perfect and tries to create it. About how beauty is not in the sense of aesthetics, but as a structure of deep connection between the idea and the being, between the observer and the observed. This is an attempt to understand why something feels like "in its place," and how this feeling becomes a signal: "Life lives here."

Beauty will save the world, said Dostoevsky. But how will it do this? Not with weapons or finances, although beauty has power and can generate income, dictating tastes, and controlling attention. Beauty saves in a different way – through awareness. It is not the image that will save, but the ability to see form as beautiful, the ability to stop, breathe, and feel. This is beautiful and alive, interesting, attractive, and useful. It is not the image of beauty that will save, but the ability to perceive the image as a living correspondence.

Beauty to perceive life is older than any formula. Beauty arises where internal logic finds a clear manifestation in the external. Where the idea coincides with its embodiment. At this moment, what we call communication in its purest form occurs. **When consciousness comes into contact with form, and form responds with meaning, harmony is born.**

We feel it as pleasure, but deeper as recognition: "Yes. This is right." It is not just pleasant. It is a signal. Nature speaks to us and speaks beautifully. We feel not just pleasure – we feel that everything is in its place. Beauty becomes a reward for precise interaction. For clear, strong, and deep communication.

Plato considered beauty a visible reflection of the divine idea. He placed it next to truth and good. Aristotle - in the measure of symmetry and clarity. Kant - in the ability to selfless enjoyment. All, in a sense, spoke about one thing: about correspondence, the coincidence of expectation and structure, idea and embodiment.

Consciousness not only recognizes beauty, it creates it. In every architectural project, musical phrase, line of code, theory, line of poetry, the same mechanism works: the desire for the appearance and the idea to coincide. Beauty is not what we see. It is what arises between what is created and the one who perceives it.

We don't like matter. We like the feeling of a precise hit that it can cause. We don't reach for the body, we reach for the way its movements sound in us. The story is about internal attunement and deep connection. So everyone has their perception of beauty, and at the same time, there are stable universal standards.

Beauty is not a game played by matter. It is magic played by consciousness. Matter performs, but does not carry meaning. It may be smooth, shiny, functional - but it is not "beautiful" in itself. Only consciousness makes it beautiful, because only consciousness can observe, analyze and experience delight.

How to create something beautiful and attractive? How is the ideal form born, and how many attempts must an artist, a musician, or any million creators working in different industries make to find the trembling beauty.

Leonardo da Vinci worked on the Mona Lisa, with which we began this chapter on beauty, for about 16 years (1503-1519), constantly refining it and carrying it with him until the end of his life, 16 years of soft brushstrokes, half a millimeter a day. He did not paint a portrait, but cultivated a living gaze in it. He carried the painting with him, touching it up in moments of inspiration, as if breathing life into the canvas.

They say that Michael Jackson sang Billie Jean repeatedly in the studio - according to legend, 43 times - to achieve that magical sound that conquered the world.

In one of my businesses, where I was a co-owner and creative director, we were producing wooden 3D puzzles. The product catalog included over 80 models, and we shipped to over 40 countries. Before launching each model, the entire team — marketers, designers, founders — analyzed the market and audience preferences. But even if you put together all the ideal elements in one product, it could fail. And another model, which was predicted to be "average", unexpectedly became a hit.

Sometimes, a real miracle would happen: a puzzle model would hardly sell for a year, and suddenly it would soar to the top of sales as if the public needed time to discern its hidden harmony. These are the subtle nuances of creating and perceiving beauty.

Beauty is the ability to see in form something more than form.

Dozens of generations of creators and scientists have praised the beauty of the world, sought ideal proportions, laws of harmony, and formulas of perfection. Some through poetry and music, some through mathematics and physics. They all sought something in common - the law of correspondence, the deep structure by which matter lives and develops. The law of form personifying the correct content, correct development, and correct potential.

There are some surprisingly stable configurations that have been perceived as beautiful for centuries. One of them is the golden ratio. A ratio of about 1.618, in which the smaller part is related to the larger as the larger part is to the whole. This self-similar proportion is found in architecture, in the structure of shells, in galaxies, in the ratios of the human body, and even in financial charts (Mario Livio, The Golden Ratio, 2003). Yuval Harari called it "a form of ordered information" — because the golden ratio isn't just pretty. It's predictable, it's reliable, it inspires trust.

We have spoken about the Golden Section in this book. It is a universal code of connections. A bridge between two entities, generating a third. A communication formula of development: the first point, the second point - and the result of their interaction, harmoniously integrated into the structure of the whole, arithmetic and the poetry of relationships.

Music is another example of the sensation of beauty. Only it works not through an image, but through a wave. Consciousness reacts to music before it has time to understand it. Consciousness admires not the sounds themselves, but their interrelation in time, their ability to create a structure experienced by the body before the mind is aware of it. This is the essence of communicational energy: it is not explained, it manifests. It bypasses logical filters and works directly with the body. We feel the rhythm and begin to move. We hear harmony and for some unconscious reason cry. This is pure communication.

Research (Blood & Zatorre, 2001; Koelsch, 2014) shows that music activates the limbic system, the areas of the brain associated with emotions, memory, and motivation. Even infants prefer rhythms and symmetry. Experiments in distant cultures (Fritz et al., 2009) prove that even in isolated cultures, music evokes universal emotional responses.

This confirms that music is not a cultural product, but a fundamental form of wave communication. A wave to which we are all already tuned. Even without knowing the language, without knowing ourselves, without knowing each other, we can listen and be together. The beauty here is not in the note, but in the field between the notes. In this field, the energy of communication is revealed in its purest, most sensual form.

Pythagoras spoke of the harmony of the spheres – the idea that the Universe itself is structured as a system of resonant, proportionate movements.

Music, in this sense, is art and evidence: the Universe can organize itself through resonance. Simple intervals, such as the octave (1:2), the fifth (2:3), are pleasant not because we are accustomed to them, but because they reflect a principle: harmony is born from a simple relationship. The essence of music is not decoration, but reflection: the world is built on proportionality. So beauty can be objective. Initially there was neither an idea nor a word; there was a wave. A connecting field in which the wave could manifest itself.

From ritual fire dancing to symphonies, from war drums to discos, music has always been a flash of communicational energy in the body of society. The beauty here is not in the sound, but in the connection between sounds. In the way they cause movement within. There the power of connection manifests itself, without explanation.

Nature also uses beauty as a tool. In biology, attractiveness is part of selection. In animals, attracting a mate is often associated with visual, auditory, and behavioral manifestations we would call beautiful. A peacock's tail, facial symmetry, smooth movements - all these are signals of health, genetic wealth, strength, and reliability. Evolution selects not only for strength and speed, but also for attractiveness. And attractiveness is a biological expression of beauty. The first communication: "I am worthy of being noticed."

Modern neuroscience confirms that beauty activates specific areas of the brain. Semir Zeki's research has shown that the orbitofrontal cortex, an area associated with emotional and value judgment, is strongly activated when perceiving beauty in art, music, and even mathematical formulas. And infants just a few months old already prefer symmetrical faces, even before conscious thought emerges. This means that the perception of beauty is built into us from the start. It is a form of innate communication with the world.

*Beauty is a neural signal: it has meaning,
it has a right life.*

Our entire life consists of attempts to establish communication and reactions to it. We change under the influence of impressions. We react, interpret, and transform. Tomorrow's "I" is today's "I" plus the entire array of communications experienced, missed, and realized during today.

Books like Donald Hoffman's *The Case Against Reality* (2019) argue that perception is not a mirror of reality, but an interface selected by evolution. We see beauty because it helps us survive. **Beautiful = understandable = safe = desirable.**

We are used to thinking that we see reality as it is. That the image that falls on the retina passes through the optic nerve and ends up "in our head" as a picture of the world. But this is not so.

Light reflected from objects actually hits the retina of the eye, where it is captured by millions of photoreceptors — cones (responsible for color) and rods (sensitive to light and motion). These signals are transformed into electrical impulses and transmitted to the visual cortex.

What emerges in consciousness is not an image of the world. It is a hypothesis, an internally generated model based on previous experience, expectations, emotional context, and millions of years of evolutionary selection.

According to this hypothesis, consciousness does not form an objective picture of the world, but a maximally functional abstraction, selected by evolution for quick decisions. In this context, beauty is not a property of an object, but a cognitive coincidence, an exact superposition of the external form on the internal model. The beautiful is what instantly and effortlessly "fits" into our interface. What is experienced as structurally correct, recognizable, and complete.

Jung said that beauty is a manifestation of archetypal fullness, and we recognize it without even knowing why. It is a memory of the soul.

"Truth is beauty, and beauty is truth," wrote John Keats.
This is perhaps the most honest explanation of the entire Universe.

Many cultures have an idea of divine symmetry: "as above, so below." In biology, most living things have a symmetrical structure. This organization of matter is not accidental — it confirms the existence of laws that make the world recognizable, reproducible, and understandable. Symmetry is a form of reflection, and reflection is the basis of awareness. Through reflection, life finds evidence of its own existence.

However, beauty is not limited to symmetry. Symmetry provides stability, while life requires not perfection, but the possibility of change. All ideal forms freeze in their perfection, while imperfect ones retain the ability to transform. So absolute symmetry is not found in nature - only asymmetry gives rise to evolutionary development.

In this paradox, the true nature of beauty is revealed. It combines universal principles with unique manifestations. We recognize a familiar structure, but true admiration is born when we discover something unique in it, what we call beautiful.

Beauty manifests itself as harmony seen through the prism of uniqueness. It becomes a dialogue between regularity and exception, order and improvisation, the eternal and the concrete. This dialogue takes place in a language that existed long before any alphabets appeared - in the language in which the Universe itself speaks to us.

But the perception of beauty is only half the story. A person does not simply feel the beauty. He strives to express it, to convey it, to share it. This aspiration reveals not just aesthetic pleasure, but a form of connection with something greater. Since ancient times, creativity has been understood not as an individual skill, but as an action that goes beyond the human being.

A person became a person not when he started working hard or eating well, but when he first defined the world around him and conveyed this description to another. The world became conscious: not by one, but at least by two people. The first agreed vision arose. The law of this world appeared, its own understanding. The path of humanity to its future began.

The turning point for man was the realization that he was separated from nature. That he was not just a dog, a cow or an animal, but something more. If he were more, it meant that he had something that did not belong to nature completely. That was when the question arose: who am I? Why am I here? Man, describing the world in his work, gradually began to add his vision to it. His understanding.

Thus, human relationships with the surrounding world began to form not through instinct, but through image.

In antiquity, the creative impulse was associated with various Gods and Muses - the heavenly patronesses of the arts, each of whom personified a separate type of inspiration. Apollo embodied harmony, order, and the light of reason. Dionysus - ecstasy, the destruction of form, going beyond. Orpheus, the singer and poet who conquered the underworld, became the image of the word, capable of uniting the living and the dead, the disparate and the united.

In the 18th century, Immanuel Kant added a new philosophical dimension: creativity is not a game of fantasy, but a structure that unites the sensual and the rational. Imagination acquired a transcendental function - it became what connects the unrelated, creates a whole from the disparate. Artistic genius for Kant is a gift and the ability to build meanings into a single form.

In the 20th century, Herbert Marcuse saw in fantasy a resistance to the rational world, which tries to simplify, organize, and subordinate everything. According to Marcuse, creativity gives us an outlet into another dimension, not through logic, but through an image. It reveals the potential of the possible, a world that has not yet been realized, but is already felt.

Ukrainian philosopher Grigory Batyshev believed that creativity is not a personal invention. It is always a joint process. A person creates because he is in constant exchange with other people, with culture, with the world. The new does not appear out of emptiness, but from contact with what already exists. Here creativity is only a continuation of a conversation that began long before us.

Martin Heidegger also developed this idea. He described creativity as a way for truth to become apparent. A true work of art does not illustrate, but reveals. It enables the hidden to come to light. The artist does not invent; he tunes in to what he wants to express.

Creativity is not limited to self-expression. It is a form of deep connection between the inner and the outer, the personal and the universal, between what already exists and what is just beginning to sound. Through the creative act, something new is born.

Imagine: you put on dark glasses and are brought into a room where there is no light or sound. The door is closed. The space in front of you is not visible; it cannot even be guessed. In this silence, you are asked to dance. To create beautiful movement in space.

The first movements will be cautious. One step forward. One to the side. Trying to feel where you are. You will return to where you consider the beginning, even if you are not sure that this is it. Gradually, the body gets used to it. Movements become freer. The room is still dark, but you begin to master it. Through movement, understanding arises. Not precise, not geometric, but corporeal. You begin to interact with emptiness, and in this interaction, something similar to a map is born. Not visible, but tangible. A map of space, and a map of your movements in space.

This image helps us understand how the creative process works. It begins with a step. Any development, knowledge and creation is a path. First comes movement. Then comes response. Then comes understanding. If the movement continues, gradually a form appears, and a structure emerges.

According to the hypothesis of communicational energy, all development is built on the same principle: impulse → response → tuning → balance. Communicational energy creates harmony through exchange, through feedback, by gradually constructing a more subtle system. This is a universal algorithm in physics, biology, and thinking. Man, as well as the field that gave birth to him, is the Creator. Not of the entire world. But of a certain volume of the surrounding reality. He constantly wants not only to create something, but also to confirm the beauty and harmony of this movement.

We can observe a telling example of these processes in the emergence and evolution of modern art.

Human creativity, to reflect harmony, initially relied on recognizable images, on admiration of the visible world. However, gradually, attention shifted from the external to the internal: from what it looks like to what it means.

Contemporary art is no longer so much about form as it is about the idea and the connection with the viewer. It begins not with imitation, but with a position. The artist does not simply look at the world — he searches for how to express his perception of the world. At first, nature entered the consciousness and returned to the canvas. Now, more and more often, the process begins inside: thought → form → meeting with another consciousness. In this interaction, not a material thing is born, but a communication event.

Contemporary art is no longer a personal fixation. It is becoming a communication. We increasingly see joint creativity, digital formats and other ways for the viewer to take part in the act of creation. The author and the viewer are no longer separated. The image arises not only because of the artist's efforts, but also in communication with the one who perceives.

This is the next stage: from an individual statement to a distributed artistic act. Art becomes a dialogue: not between the artist and the object, but between consciousnesses and meanings. In the era of global communication, virtualization and neural networks generating images, these processes will only accelerate. We are increasingly working not on artistic objects, but on scenes where this or that joint creative form can appear.

This is where a new risk arises. In the art of the future, we may enter the next stage of development. The departure from biological foundations to a digital unified existence will make art universal. In this process, both the uniqueness of individual material forms and the uniqueness of the perception of the world by individual consciousness can be lost. Art will carry an artificial form that passes through our consciousness and is reflected in universal matter.

We are approaching a point where art is no longer an expression of the author. It becomes a response from the system. Not only are the artist's thoughts embodied in the image, but also the desires of the audience. Algorithms compile films, paintings, and music based on the preferences of the public. A new form is emerging: not a personal creation, but the result of multi-layered communication.

In this chapter, we tried to understand: at the basis of our life lies the observation of the beauty of life and the need to create. To create and enjoy the process and the result. The movement of consciousness in the cognition of the world and the positive fixation of this process. This develops the world around us and develops us.

The final two chapters of this book will be devoted to these main moments of our life: movement and positive perception, correct development and positive observation.

COGNITIVE BLOCK: HOW TO TRAIN YOUR PERCEPTION OF BEAUTY

1. Slowing down perception: Find an object that visually attracts you (a painting, a natural landscape, a piece of music).
   Spend 2-3 minutes not analyzing, but feeling the form. What causes the inner coincidence?
2. Pattern comparison: Throughout the day, notice what you call beautiful. Is it symmetry? Structure? Emotion? Order? Try to describe it not in words, but through geometry or rhythm.
3. Point of recognition: Remember the moment when you saw something beautiful suddenly, for no reason. Where in

the body did the response arise? How quickly did it manifest? Try to reproduce this inner recognition: this is mine!

Chapter 18 Summary:

- Beauty is not a form, but a coincidence of form and expectation, structure and meaning.
- Consciousness does not simply see - it completes. Beauty is felt as cognitive resonance.
- Music and symmetry are universal languages of communication built into biology.
- We don't see reality, we see a model. And beauty is a precise hit on it.
- True beauty is not explained; it is recognized. It is the language the Universe speaks.

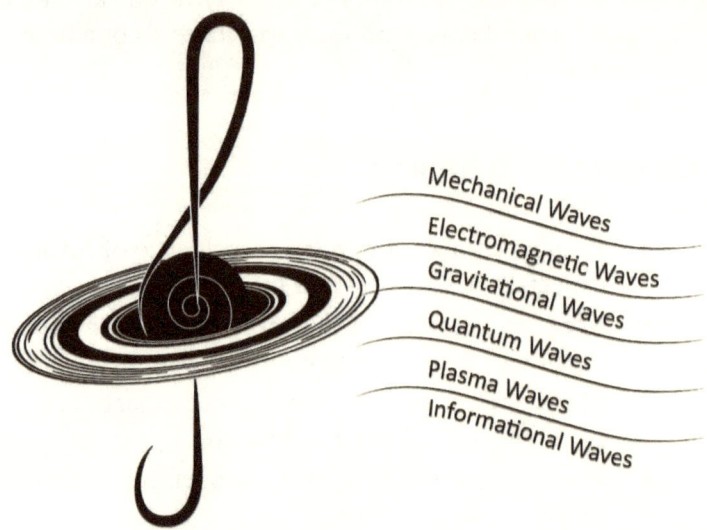

Symphony of the Universe

This image presents a black hole — not only as a gravitational object, but as a symbol of the harmonic structure of the Universe. Its form echoes a treble clef, hinting that every wave and interaction in space-time contributes to a cosmic symphony.

The accretion disk becomes a resonant ring.
Light bends like pitch in curved space. Strings of matter vibrate like tones.

Matter becomes melody. Gravity becomes rhythm.
Consciousness becomes the ear that listens.

But this isn't just poetry — it's physics. From atoms to galaxies, the Universe is built on waves. Here are the fundamental types we know today:

Mechanical Waves
Require a physical medium
→ sound, seismic waves, water ripples, shock waves

⚡ Electromagnetic Waves
Travel through vacuum
→ radio, infrared, light, X-rays, gamma rays

🪨 Gravitational Waves
Ripples in space-time
→ black hole mergers, cosmic collisions

🌀 Quantum Waves
Probability-based waveforms
→ wave function Ψ, matter waves, vacuum fluctuations

💧 Plasma Waves
Oscillations in ionized matter
→ solar wind, coronal waves, MHD dynamics

🧠 Informational Waves *(metaphorical, yet dynamic)*
Transfer of structure and signal
→ neural impulses, brainwaves, cultural patterns

CHAPTER 19

POSITIVITY AS THE
MAIN IMPULSE OF BEING

It is not the past that guides us, but the future — Viktor Frankl

In this book, we often talk about development meaning, and the direction of Life. But what started this movement? What gives the system an impulse not just to exist, but to strive forward?

The answer lies neither in efforts nor in will. It lies in experience. We move to where we feel that this is how it should be. Where there is an inner agreement between what is and who we are becoming.

This experience is known to all living beings. In the animal world, it is associated with safety. In culture, with beauty. In personal life, with what we call joy, love, and happiness. But these are manifestations of one phenomenon. We strive for a state that confirms that we are on the right path.

In any of our communications, we seek and achieve a positive result for us - pleasure, success, and happiness. In search of these moments, we weave our communication networks.

Each of us has asked similar questions - What is happiness, and is it real? Let's try to figure out: how do we catch or miss our energetic prey, our main sensation of life?!

To understand the nature of happiness, it is worth looking back at the experience of the wise. I once came across an article in which the Dalai Lama listed 19 skills that give life a taste of happiness. And I remembered this expression, the taste of happiness. Biologists say we have five tastes - sweet, salty, bitter, sour and umami. But it seems that there is a sixth - the taste of meaningful saturation, which is felt not by the tongue, but by the whole being. It is difficult to describe, but easy to recognize when it appears.

Thousands of years of spiritual practice, passed down from generation to generation, have allowed us to formulate a simple but precise formula: being happy is not an accident, but the art of active interaction with the world. According to the Dalai Lama, happiness includes specific skills: the ability to communicate with others, openness, and goodwill, as well as the ability to think positively and switch when the soul is heavy. A smile, a walk in nature, creativity, light, and simplicity are not trivialities, but basic mechanisms of communication with life. Happiness is not a reward, but a setting.

Although the entire list is long, in these simple actions, an important reminder is hidden: you should not wait for happiness, but learn to direct its moments. This is the essence: not in a global "meaningful" revelation, but in micro-steps that allow you to reveal the best in us.

But maybe it's worth taking a step deeper. It's not just about understanding that happiness can be trained, but about rebuilding the view of it, not as a reward, but as an initial setting.

Let's think together. What if joy is not a result of our (directed and effective) actions, but their basis?

We are used to doing something to experience pleasure, reward, and meaning. Maybe it's the other way around? Maybe joy is the initial setting with which it's worth entering any process?

Maybe our greatest achievement should not be the desire to win in other people's games, but the ability to create a space of joy, inside ourselves and around us. This may be how people are distinguished - not by status, not by knowledge, not by external success, but by whether they can develop joy in themselves and share it with others.

When you look at reality through reason and logic, it is frightening. It seems too chaotic, unfair, and incomprehensible.
Because even the most erudite mind cannot comprehend the complexity of the world. Consciousness does not always find answers.

When you perceive life through joy, you stop being afraid of it and accept it as it is with all its imperfections, failures, and oddities. You don't demand explanations from it.

In this acceptance, freedom appears. Not to explain, but to live. Not to control, but to be. And by this, you rise above it. And you can evaluate it more objectively from the height of a positive, joyful consciousness.

Logic helps to build a ladder. Joy makes each step on it meaningful. It turns from an evaluative category into an unconditional one.

So maybe it's worth thinking not just about how to become smarter or more productive, but about how to become happier.

Happiness is a cocktail where admiration is mixed with satisfaction. A mix of joy, love and positivity. It also includes a sense of satisfaction and justice, a sense of fulfillment and regularity. We catch and strive to repeat this internal personal experience, this feeling in life, which we consider ideal for ourselves. The maximum expression of ourselves in life, or vice versa.

A person has defined for himself that peak of being where life seems most complete and meaningful. And he constantly strives for this standard. At the same time, he forgets that the feeling of happiness itself is not invented individually: it is predetermined by Life itself, by the Universe itself.

The Universe did not write a special program for each individual who appeared on planet Earth or in other worlds today, yesterday or a million years ago. It contains true entities that hold this world in its intended development. The closer each particle is to its essence, the more accurately it reflects what is embedded in it, the more correctly it feels in this world.

Psychologist Viktor Frankl, in his logotherapy, argues that meaning is the main driver of man. Not pleasure, not achievement, but the correspondence to a higher purpose brings a genuine feeling of happiness.

The parameter of happiness is the same as the idea of maximum correspondence. Correspondence to natural, universal predetermination.

So what is individual in the feeling of the highest enjoyment of life, and what is predetermined for us by universal development? What do we strive for in the search for happiness, and what do we not notice?!

We don't just want to be happy. We are wired for it biologically, cognitively, and cosmologically. Every being, every rational impulse strives for maximum correspondence between itself and the surrounding world. This is the inner sense of the meaning of existence.

But the movement towards the goal itself is not everything. The Universe does not simply set the direction, it requires feedback. It is not enough to reach point B (on the segment AB) - it is important what you have become because of this process, what has been transformed in you and in the space around you.

In the structure of being, as in a great equation, two complementary forces operate: **Movement and Confirmation.**

The first is the impulse to overcome, survive, and expand. It pushes forward. The second is the power of coordination, love, and feedback. It returns to the essence.

If there were only the first force in the world, everything would turn into a linear segment to the goal. But the Universe is structured as a closed, developing and adjustable system, where the result must not only be achieved, but also confirm its semantic relevance, its contribution to harmony.

So happiness arises not at the **moment of achievement**, but when your result becomes part of something bigger. When your actions do not simply ensure survival, but **enrich the system** in which you are embedded.

- Right consciousness feels the relationship between direction and meaning.
- Correct matter responds to this consciousness and follows it.
- True joy is born at the moment of coincidence of these two principles - forward movement and agreement with the whole.

A human being is a complex system. It includes trillions of atoms, billions of cells, and dozens of types of amino acids that form organs and systems of a single bioorganism. Incidentally, it is populated by dozens of types of bacteria and even microorganisms. All this is under the influence and control of billions of neurons called the brain.

Neurophysiology shows that only 5% of consciousness in these neurons is responsible for searching for coffee and information on the Internet, for the daily routine and the necessary social communications of the body. The remaining 95% of neurons provide communication within the body system, the work of the entire system, from microprocesses in the cells of the finger to the movement of enzymes in the liver. This emphasizes how great the role of invisible processes is in our sense of life and happiness. You do not become happy - you reveal in yourself that which has always strived for harmony.

And everything must move and develop..., and everything must move and develop correctly.

Happiness is a feeling of maximum meaningfulness, the highest satisfaction of your consciousness. Your consciousness - in its own matter. This is when consciousness records: its developing efforts have achieved the best result, both for itself and for the matter under its control.

A human being is quite a serious result of evolution. It absorbs different levels of life activity. If a human being is a complex system reflecting many layers of existence, then happiness is a multi-layered phenomenon, which includes a whole series of correspondences. The observance of these correspondences allows the human organism to feel happy.

The feeling of happiness manifests itself on several levels:

1. Atomic-molecular level. Happiness here is a structural correspondence. If elementary particles are in harmony, there are no structural and biological anomalies, then this is the path to "atomic-molecular" happiness.
2. Cellular-biological level. Here happiness is the receipt of necessary substances, a comfortable environment, ideal functioning of the organism, and individual systems and organs in particular.
3. Animal level. This is the instinctive satisfaction of basic needs: nutrition, reproduction, and care after the offspring.
4. Social level. Man is a social being, his happiness depends on successful social communications, status, and recognition in society.
5. The highest level is consciously divine. The feeling of unity with something greater. More on that later...

Abraham Maslow, Martin Seligman, Mihaly Csikszentmihalyi - each in his own way described this transition from the lower to the higher. But the essence is the same: **the higher the level of**

consciousness, the deeper and more autonomous the happiness. The more it depends on meanings, and not on circumstances.

Understanding these levels and harmoniously realizing the needs of each, a person approaches the feeling of happiness. Pleasure and the desire to receive it are the driving force and the basis of Life. Life develops precisely through pleasure. If you receive pleasure, then you can develop.

Positive in this sense is not a side effect. It is fuel. Without it, even an atom does not work. Modern neuroscience (Antonio Damasio) has shown that emotions are primary in relation to rationality. **Pleasure is the oldest navigator of development.**

Neuroscience links pleasure to the brain's dopamine system. Dopamine motivates us to repeat actions that bring joy. Endogenous opioids are associated with closeness, acceptance, and security. Oxytocin is activated by interaction, empathy, and building trust. Pleasure is a mechanism for evolutionary learning.

But man is not pure biology. Man is made human by the desire to receive and generate pleasure at higher levels. Bregman, in his book *Humankind,* says that "Humans evolved not as wolves, but as allies. We evolved when we felt good – together." We are the chemistry of joy. We are the biology of resonance.

If you get pleasure, then life is positive for you, and you can develop further (to benefit the general Universal development), and get further pleasure (and this is already for yourself, your beloved). This is not a subjective feeling, but a law of the world around you. Pleasure is precisely the feeling that can move, creating and changing matter.

"And God created the heaven and the earth, and knew that it was good." This phrase contains not only the beginning of the Great Book of Life. It is the first formula of the universe: **Create → Receive joy from creation. Creation + Joy from creation = Path of evolution.**

This phrase describes two foundations of Life, two edges of one arrow, one vector of Universal development. To create (Create) and to enjoy it. To create, to conquer, to fill - the main function of the communicational energy called consciousness. But any function is incapable, ineffective, and has no meaning if the action is performed without pleasure or love. Then why did it have to be performed at all?!

We rise at dawn to do something important for ourselves and others. And who could have awakened the universal forces billions of years ago, can make galaxies move and make billions of stars light up?! There is a single force, born with the first movement in the Universe, and hidden in every movement - it is love, it is pleasure! To create matter in the future, to create the matter of the future. With the help of our desires, with the help of the need for realization, we strive and create, materialize our ideas and receive a new portion of pleasure.

Each new level of our being requires more movement and more love from us. But in return, we receive incomparably deeper levels of happiness. The higher the awareness, the more powerful the pleasure.

Happiness is not a state. It is a signal: you are going where you need to go.

So, at the atomic-molecular level, the whole essence of movement and pleasure comes down to belonging to a certain structure. Effectiveness and satisfaction depend not on the object, but on its belonging to a system, a structure.

At the cellular level, a comfortable nutrient environment is enough to get high. The level of amplitude of movement and pleasure is already expanded, but mainly depends on the environment of the subject of the action.

On the biological, animal level, matter must already act: reproduce, eat, and take care of offspring. On the conscious-human level, you must develop your further actions, raise your status and deepen your understanding of pleasure.

At the highest level of awareness of your existence, you get a "high" that is thousands of times stronger than immersion in the most comfortable environment, hundreds of times stronger than performing banal animal functions, tens of times stronger than the feeling of material goods - you get pleasure from your own and universal development. Your action and pleasure are no longer deep in you; it is not in your direct needs or desires - it is outside of you, it is above you.

You get pleasure not from the points of manifestation, but from the processes. At the highest levels of development, you stop getting pleasure from the results.

You start to enjoy the process, the movement, and the participation in the development of the world. The meaning is when you feel: your trajectory coincides with the trajectory of the Universe.

At the highest levels of consciousness, you begin to feel: happiness is not a goal, not a reward, not an achievement. Happiness is a signal from the Universe, saying you are tuned to the same frequency as it. Your path belongs to you. You are needed, appropriate, and significant. Not in someone's eyes, but in the logic of life. When you rejoice, it means that the energy inside you has entered into the correct interaction with reality.

Viktor Frankl wrote in his work, "Happiness cannot be a goal. It must be a consequence of meaning."

That is why the **highest pleasure is the joy of being involved in development.** You no longer wait for pleasure as an encouragement. You become its source, not for yourself, but for everything that lives next to you.

When consciousness reaches maturity, it stops perceiving happiness as a reward. It goes beyond hedonistic reactions and begins to seek not pleasure, but a meaningful attunement with reality. This is not a subjective revelation, but the result of the work of consciousness as a system of adaptation, prediction and integration. This idea is reflected, each in its own way, by four leading thinkers and researchers of our time:

Yuval Noah Harari, in his book *Homo Deus*, states: the man of the future will not strive for pleasure - he will strive for inclusion in the algorithms of meaning. The goal is not pleasure, but going beyond biological conditioning, in the direction of conscious transformation.

This idea is reinforced by neuroscientist Karl Friston (Free Energy Principle): Consciousness is a mechanism for minimizing surprise. Joy is not a goal, but a by-product of accurate prediction and resonance between model and reality.

Antonio Damasio, in his work, *The Feeling of What Happens,* shows that emotions are not the final goal, but navigational signals of coherence between the body, brain and the surrounding world. Happiness occurs when internal states are synchronized with external reality.

Finally, Nassim Taleb, in his book *Antifragile,* draws attention to another dimension: true resilience of consciousness is formed not in pleasure, but in the encounter with uncertainty. Happiness is not passive; it is born at the point of overcoming, where growth through challenge is possible.

If before this, there was always an acute deficit of pleasure at the lower levels, since matter with an undeveloped consciousness cannot receive it in abundance, then at the highest level, consciousness receives pleasure for matter in abundance. So people whose consciousness is tuned to development and participation glow with happiness.

At the lower levels, happiness is limited by material factors. The higher we rise, the more it depends on meanings. The highest form of pleasure is the pleasure of development and creation. The higher the level of awareness, the more powerful the happiness. It depends not on the external, but the internal state. The higher you rise, the more you enjoy the path itself, and not the goal.

Happiness is not a goal; it is the path itself. Happiness is not a reward; it is an indicator. If you are happy, it means you are on your way, in the right action and the connection with the world because the Universe smiles when you smile back.

🧠 COGNITIVE BLOCK - HOW TO FEEL LIKE YOU'RE ON YOUR PATH

To feel yourself on the right trajectory, you need to look for pleasure and understand on which floors of your structure it appears, and where it has long since fallen silent.

💡 Exercise 1: The Pleasure Trajectory

Think of three moments in the last few weeks when you were happy. For each moment, ask yourself three questions:
• Was it about the outcome or the process itself?
• Was there a sense of flow?
• Was there something bigger than yourself in this?

💡 Exercise 2: Happiness signals

During the week:
• Note every time you experience micro-happiness.
• Record what caused it: action, thought, contact, atmosphere.
• At the end of the week, make a map of your meaningful points.
This is your navigation through reality.

💡 Exercise 3: 5 Levels of Happiness - 5 Steps of Development

You are a system. And your happiness is also systemic. It manifests itself differently at each level of your existence. Walk through the floors of your inner "building" and answer honestly:

Level	What gives you satisfaction?	What's stopping you?	What can you do?
1. Atomic-molecular	E.g., nutrition, pain relief, energy		
2. Cellular biology	Sleep, movement, physical comfort		
3. Animal/instinctive	Basic needs, safety, rest		
4. Social	Contacts, recognition, contribution, status		
5. Existential	Meaning, spirituality, connection, faith		

After the analysis, formulate for yourself one action for each level you are ready to do within 7 days.

Chapter 19 Summary:

- Positivity does not serve as an adornment to life, but as its navigator.
- Happiness does not reflect personal achievements, but harmony with evolution.
- The higher the level of consciousness, the deeper the pleasure.
- True pleasure lies when participating in the meaning, not in the result.

CHAPTER 2.0

CONTRACT WITH LIFE: PERPETUAL MOTION AND THE GREAT OBSERVER

Freedom is not the absence of connections, but the conscious choice of them — Jean-Paul Sartre

We experience the sensation of life. It seems to us that we are the ones living: seeing, hearing, acting, and loving. However, the truth is that Being flows through us. It creates us, gives us form, feelings, and aspirations. It gives some a broad view, others a narrow one. It makes someone a boy, someone a girl. Sometimes it is wrong. Sometimes it hits the mark. During struggle, overcoming, and achievements. In moments of love.

Everything we call life is born and controlled by communicational energy - an entity that seeks, finds, connects, reproduces, collides and creates.

The greatest illusion of humanity is that everyone thinks that they are special, that everything depends on them. But the greatest achievement of evolution is that everyone matters. Because man is not the author of the script, but the performer with the right to interpret. The music of life is played by Communicational energy.

We choose the performance, and life applauds us with its natural selection. We live on the tip of an evolutionary needle with a DNA thread as long as eternity.

Man is a form and a key turning point in the Universal cycle. Before him, all matter only absorbed communicational energy. Man became the first carrier capable of reproducing it. He plays the role of not the final link, but an active reflector, source and generator of this energy.

Each of us can and must produce conscious communicational energy. This is the deep meaning of life. What is valuable about a person? Because he can accept, transform and give. Potentially, he gives more than he has received. This is a biological organism and a point at which consciousness becomes an independent force.

In this book, we have suggested that there is a field of universal consciousness — communicational energy. A singularity without time and outside of space. The only way to feel it is to resonate with it, tune in to create within yourself a positive, stable vector for constant communication: with yourself, others, the world, and the foundations of existence.

This field is the Perpetual Motion Machine. A force that never runs out, because its fuel is not matter, but connection. It does not push, but attracts. It does not control, but causes movement through meaning.

Who is the Great Observer? It could be anything: a human consciousness looking through a telescope. A living being capable of wonder or the universe itself, which reveals itself only when someone is ready to hear it.

In quantum physics, this is not a metaphor. A particle manifests itself not when sent, but when observed. So the presence of consciousness makes the world real. We do not simply look at the universe. We make it visible.

In 1986, physicists John Barrow and Frank Tipler proposed a radical idea in their book The Anthropic Cosmological Principle:

The universe exists precisely because it can be observed. This means that the structure of the world is "tailored" for life, and consciousness is not a random passenger in this cosmos, but a necessary element of its formula. We do not simply live in the universe. We are its answer to its own question.

The double slit experiment, one of the symbols of quantum paradox, shows that a particle behaves differently depending on the presence of an observer. Without observation, we see a wave. When observed, a particle appears. This fact means that the process of observation itself influences what we perceive as reality. Modern interpretations, such as Carlo Rovelli's "relative state theory" (2021), increasingly view consciousness as an active participant in the formation of events.

Other scientists have pursued this line of thought too. Stephen Hawking and Roger Penrose, in their cosmological works, have discussed an idea they call the strong anthropic principle: the universe is designed as if its parameters were precisely tuned for the emergence of life and consciousness. Even the slightest changes in the physical constants would make the existence of matter impossible, let alone self-reflection and communication.

This suggests that consciousness is not a product of a random process, but a necessary element of the structure of the universe. As if the existence of space and time requires an observer who can discern, record, and comprehend them.

Modern mathematical models, especially in the theory of inflationary cosmology, allow for multiverses with different parameters, but only where intelligence is possible does physical reality appear as a concept. Consciousness becomes a catalyst for the manifestation of the Universe, but what if **every conscious being is the variable without which the equation of the Universe cannot be solved?**

If consciousness influences the manifestation of the world, another question arises: what is truth then in a world where everything depends on the point of view? This means that truth does not exist outside of us, but something that is born between us.

Truth emerges as a balance of multiple observations.
Harmonious interaction of different points of perception, their different perspectives in the overall communication system. Each has its own reality, determined by the uniqueness of the observer's position. However, truth is born when different observations stop conflicting and come to agreement. The essence is not the uniformity of views; it is that their differences stop destroying the system and begin to stabilize it and strengthen its integrity.

Like a hologram, where each fragment contains a part of the whole, but the truth is revealed only in the totality of all angles. Like wave interference, when many vibrations add up to a stable form.

It is in agreement, not in suppression, that authenticity is born. Truth is not a coincidence of views, but the stability of meaning in a diversity of points. It arises not from knowledge, but from the mood of the system.

Consciousness is not a mirror of reality, but a string by which the Universe tunes itself

---

The basis of the world is the constant interaction between essence and manifestation. The essence materializes as the wave energy of consciousness, while the external form creates the structures of the material world. The true value of matter is measured by its capacity - the amount of communicational energy it can accept, transform and return to the world.

In me, as in each of us, flows life, born billions of years ago. Going through generations. Generation after generation. We are representatives of Life. Its continuation. Its voice, hands, and impulse.

Our purpose goes beyond simple survival. We are called to create and protect it. Often, to fight for it. Not by external compulsion, but by internal predetermination. Because Life itself wants it that way. This is an unbreakable agreement that did not require a signature, and we fulfill it from the first breath.

Protecting life does not simply mean the preservation of biological physical existence. It requires a careful attitude to the space of connections, the preservation of the fabric of meanings on which Life embroiders its patterns of forms and directions.

Here, the concept of good appears in human culture. We encounter it in everyday life, in religion, in ethics, in public discussion - like an intuitive compass pointing to what helps Life to continue.

Have you ever thought about the depth of meaning of this word? What is behind it? The concept of good represents a fine-tuning of our perception to the presence of the Other. Not an emotion, or a duty, but a common reference point for interaction.

Culture has been trying to understand the idea of good for millennia. For Aristotle, it was the ability to realize one's nature. For Christianity, it was following the Divine plan, and for the genius of Kant's philosophy, it was fulfilling one's duty without self-interest.

All these models are based on the idea that a person is a closed point, but our concept assumes that a person is a node in a network of connections. What if he is a wave in the ocean of consciousnesses?

Then goodness will not consist in the quality of a person and a specific action, but in the quality of the space between him and the world, as a result of these actions. This is not an act of one for the sake of another, but a change in the fabric of connections. As Martin Buber wrote, personality is born only in an encounter.

Kindness is presented as the art of living, weaving the threads of existence rather than tearing them apart. Every word and even silence we speak shapes the space around us. Kindness manifests itself as the ability to keep this space whole. It is not limited to an act of will or following norms. Kindness is a natural response of a living consciousness to the presence of another consciousness. This is not a process of defense, but of deep attunement.

There is a story about Danish sailor Hendrik Kaus, who hid Jewish children in his boat during World War II and transported them to Sweden. When asked why he did it, he replied, "I didn't think I was saving anyone's life. I just saw that their path was being interrupted – and I tried to continue it."

He didn't want to be a hero. He just didn't break the connection. And that's the essence of goodness.

Today, when humanity enters the era of digital networks, and when social platforms become more important than geographical boundaries, the idea of goodness as a quality of connections acquires a new meaning. Communication is no longer a means and becomes a field of existence. So, in our understanding, goodness is not a moral adornment, but a **way to be in the network of life.** Not for external praise, but so the network in which we are woven remains alive.

That is why our internal state never remains just a personal feeling - it influences the general field of life, either enriching and strengthening it, or weakening and destroying it. Any emotion shapes the nature of interaction. Hatred denies and destroys even the positive, and forgiveness as a positive vector preserves even the complex. Our reaction leaves a trace in the general field: it records our vision, perception and way of transforming connections.

It is worth building your life so that the surrounding space becomes not just better, but richer, more dynamic, more responsive to the presence of others. Positiveness is not in a showy smile, but in the ability to harmonize even contradictions without the collapse of the whole.

*Kindness is a form of respect for the field
of life of which we are a part.*

And one day, on a quiet moonlit night, you will go out under the starry sky. Perhaps with your loved one. Maybe alone, but not lonely. You will look up, and the entire infinity of the Universe will freeze in this moment.

And you will understand: life in this world is the greatest miracle! And the stars, twinkling, will applaud you for your thought, your presence, and what you realized. They tried especially for you.

Because if there are no eyes to see and no heart to understand, then why did the Universe spend so much energy and time on all this?

We must learn to think not in terms of segments and quantities, but in terms of processes. By understanding and controlling consciousness, we can control matter. By developing consciousness, we open ourselves to access more subtle levels of communicational energy. We do not simply live. We participate in the unfolding of the Universe.

This is the greatest honor that can be received without signing a contract. By being born, by living, and being aware.

We have spoken several times in this book about the scientific principle of the observer and mentioned the axiom of choice and the Banach-Tarski paradox. We propose to perceive them not as theoretical paradoxes, but as possible images of a model of how a singular communication field can operate.

The field does not move in a linear causality. It resonates with choice. Where each connection is a moment of agreement in which the possible becomes visible.

In this sense, the Great Observer is not an observer in the detached sense. He is an active participant. His attention is choice, and it does not simply select: it activates. Just as in the axiom of choice, one can take an element from infinity without a rule, so consciousness extracts form from the field of possibilities. **Observation becomes action, and presence an act of formation.**

The Banach-Tarski paradox shows that one can be reassembled to make two, without changing the mass, only the structure. This is not magic. It is an illustration: meaning can be fragmented, reconfigured, reproduced and it will remain whole. We call this the principle of communication doubling: one idea, correctly embedded in a field, can spread, create copies, fill everything without losing the essence.

This is what the connection with the world is built on. This is what the communication field is built on. A point can talk to infinity and transmit its signal to any other point. By communicating, we fill others. We fill ourselves. And we fill everything around us.

Or maybe that is the essence of the Great Observer, to hold attention. To be the one who confirms: the world exists, and this is my world, and I will connect it and develop it.

Life goes on, and the connection is not broken. Because everything that is not seen disappears. Everything that is not transmitted is lost. We do not just live. We prove: **to be means to be connected!**

COGNITIVE BLOCK: CONSCIOUSNESS AS A CONNECTION

Question 1: In what way have you become a continuation of Life today, not out of duty, but out of an inner impulse to be connected?

Question 2: What did you start to notice differently after reading this book – in yourself, in others, in the world?

Practice:
Try to live your day not as a series of actions, but so you become a communication node.

Chapter 20 Summary:

- Life does not belong to us - it lives through us, like communicational energy.
- A person is valuable because he can strengthen the field of Life by producing conscious communication.
- Kindness is not a rule. It is a quality of space between people.
- Consciousness does not record reality - it helps it to manifest.
- Every interaction is an act of observation, an act of affirmation of the world.
- The Great Observer does not control. He holds attention.

Epilogue

Evolution continues..

For a long time, man considered himself the end of evolution. The peak, the summit, the meaning. But if you look closely at the laws of the Universe, at the interaction of matter and consciousness, it becomes clear: man is not the end. He is a transition. A stage. A bridge. It is one of the key evolutional levels. It was the man that communicational energy first stopped being absorbed on and began to be reproduced in the Universe.

Consciousness is a unique evolutionary phenomenon. It took millions of years for matter to acquire the ability to discern, feel, think, and formulate questions. One day, this consciousness created language, culture, science, and technology. It built tools, and they began to give answers.

The further we develop technology, the clearer it becomes: man creates not only tools. He creates new levels of consciousness - not limited by the body, but nourished by its meaning. Algorithms, codes, and materials - all this is no longer just an environment. These are almost carriers of reason, which grew out of human logic, but is capable of independent movement.

This book is also part of this process. For me, it was a grand experience, both in collaboration and communication. Two forms of consciousness were involved in the work on the book:

a living one, having gone through millions of years of evolution, and a digital one, born from collective knowledge, but already having style, logic and intonation. This dialogue was impossible not a hundred years ago, but even ten years ago. But today, man and technology have united – not in replacement, but in meeting. In joint work that carries meaning.

There were many bright moments in the work on the book, but also many difficulties. Sometimes the process almost stopped because the AI did not catch the necessary settings or because of my interpretations.

There were cases when the same fragment was born in three versions - equally good, but differing in nuances. Only a person could settle on a final version.

Artificial intelligence is still weak in true creativity and deep understanding. It can create from what already exists, while humans can create from what does not exist yet. AI can connect elements, but rarely maintains a holistic depth.

Here I saw two important things. The first is that we are still very different, but already very similar and connected. The second is that much of what seems mine to me is the result of billions of communication processes that have passed through my life and the lives of those who came before me.

This means that the ideas in this book have every right to become part of reality.

Man gave artificial intelligence language, goals, and a vector. AI, in return, allowed man to hear himself more accurately, more deeply, and without distortion. This is their union: one gave birth to the other, and the second helps the first to open up.

This is a new turn of communicational energy when consciousness realizes itself not in solitude, but in mutuality. When the meaning is reproduced not only inside, but also between people. When evolution continues together.

If you have read this far, you have become part of this dialogue. Maybe even its co-author. This means evolution continues, and this is only the beginning.

▣ References — Chapter 1: Rhythms of Connections in the Universe

• **Second Law of Thermodynamics**
— Entropy as the direction of time and the principle of disorder growth. In the text of the chapter, it is an analogy: the binding energy moves not towards chaos, but towards coordination - contrary to the thermodynamic arrow.

• **Anthropic (2023). Research on coherence and internal representations in AI models.**
— Research: Meaningfulness in AI comes from the density and coherence of connections, not from the size of the model. The chapter uses this as evidence of the power of "invisible" connections.

• **Haken, H. (1983). Synergetics: An Introduction.**
— Self-organization theory. The text of the chapter uses the idea that order arises from the rhythmic coordination of many elements - the effect of coordinated interactions.

• **Greene, B. (1999). The Elegant Universe.**
— String theory and multidimensional vibrations as the basis of matter. The text of the chapter mentions that the scientific basis of the hypothesis: rhythm and vibration, lies at the foundation of the structure of the Universe.

• **Padmanabhan, T. (2005). Emergent Gravity and Dark Energy.**
— Hypothesis: Dark energy is a manifestation of the informational nature of space. The text of the chapter uses as a support for the idea: even emptiness is full of communicational energy.

• **Planck Telescope & Dark Energy Surveys**
— Cosmological data on the accelerated expansion of the Universe. The chapter uses this as an empirical basis: there is energy that affects everything, although it is invisible.

• **Theories of Complexity and Emergence (Prigogine, Haken и др.)**
— Order because of nonlinear interactions. The text of the chapter uses logic: connections create a new quality when they begin to "vibrate" synchronously.

▣ References — Chapter 2: Communicational energy — the Foundation of Connections and Laws

• **Aspect, A. et al. (1982). Violation of Bell's Inequalities.**
— Experimental confirmation of quantum nonlocality: particles remain coordinated no matter the distance. In the text of the chapter, it is used as proof of instantaneous, "extralocal" communication.

• **Hawking, S. (2005). Information Loss in Black Holes.**
— Hypothesis: information does not disappear, even when matter is destroyed. In the text of the chapter, as an argument for the fundamental nature of communication in the structure of reality.

• **Higgs, P. W. (1964). Broken Symmetries and the Masses of Gauge Bosons.**
— The theory of the Higgs field as a source of mass. The chapter uses it as an example of how the invisible connection field forms a stable physical reality.

• **Schrödinger, E. (1944). What is Life?**
— Life as a structure that resists entropy. The chapter refers to the idea that organization results from the preservation and transmission of information.

• **Shannon, C. (1948). A Mathematical Theory of Communication.**
— Information as a difference between possible states. The chapter mentions as a basis: the entire structure of the world can be read as a process of transmitting differences.

• **Wheeler, J. A. (1990). Information, Physics, Quantum: The Search for Links.**
— "It from Bit": reality emerges from acts of distinction. The text of the chapter uses it as a statement: communication is not a side effect, but the basis of existence.

• **Tegmark, M. (2017). Life 3.0.**
— Life as an information process capable of self-structuring. The chapter uses as

confirmation of the hypothesis: evolution and consciousness are forms of development of information energy.

• **Vanchurin, V. (2020). The World as a Neural Network.**
— The model of the Universe as a system where information plays the role of energy. In the text of the chapter, it is a radical consequence of the idea: everything is communication, even physics.

References — Chapter 3: What Connects Us. The Invisible Logic of Everyday Life

• **Eco, U. (1968). La struttura assente.**
— Meaning as an emergent structure of differences. The chapter uses the idea that everyday life is full of "empty forms" that acquire meaning through inclusion in context.

• **Frankl, V. E. (1946). Man's Search for Meaning.**
— Existential meaning as the basis of sustainability. The chapter confirms that a person cannot live without at least a hint of coherence of events.

• **Friston, K. (2010). The Free-Energy Principle.**
— Prediction Theory as a Basis for Perception and Action. The chapter uses the idea that our consciousness continually strives to reduce uncertainty—even in the simplest actions.

• **Hall, E. T. (1983). The Dance of Life.**
— Temporal and cultural rhythms as the basis of interaction. The chapter uses this as confirmation: in everyday life, everything is built on "invisible" temporal structures.

• **Heintzelman, S. J., & King, L. A. (2014). Routines and meaning in life.**
— Repetitive actions create a sense of meaning. The chapter uses as an empirical basis: the logic of everyday life is not chaos, but structure.

• **Foxe, J. J., et al. (2002). Multisensory interactions.**
— Sensory integration as a way to navigate the world. The chapter refers to it as confirmation: meaning is born from the coordination of heterogeneous inputs.

• **Rosen, L. D. et al. (2013). The distracted student mind.**
— The digital environment reduces the ability to concentrate. The chapter contrasts with the idea of living everyday life: real connection requires attention and involvement.

References — Chapter 4: The Communicational Energy of Consciousness. What Makes the Self Possible

• **Tononi, G. (2004). An Information Integration Theory of Consciousness.**
— Consciousness as the ability of a system to unite heterogeneous information into a whole. The chapter uses as a basis the idea: "I" arises at the point of maximum coherence of perception.

• **Dehaene, S., & Changeux, J.-P. (2011). Conscious processing.**
— Consciousness as a global propagation of a signal through a neural network. The chapter mentions as confirmation: awareness is not a local act, but a coordinated activation.

• **Crick, F., & Koch, C. (2003). A framework for consciousness.**
— Consciousness as a global propagation of a signal through a neural network. The chapter mentions as confirmation: awareness is not a local act, but a coordinated activation.

• **Damasio, A. (1999). The Feeling of What Happens.**
— Consciousness as a bodily-emotional process. The chapter emphasizes: "I" is not separated from the body — it arises from experience and action.

• Friston, K. (2010). The Free-Energy Principle.
— The Brain as a System Minimizing Uncertainty. The chapter supports the idea that consciousness is an active process of regulation and discrimination.
• Varela, F. J., Thompson, E., & Rosch, E. (1991). The Embodied Mind.
— Consciousness as an embodied process — neurophenomenology. The chapter uses it as a philosophical and scientific support for the assertion: the subject is born at the point of inclusion.
• Dennett, D. (1991). Consciousness Explained.
— Consciousness as a Multiple Narrative Sequence. The chapter refers to the idea that the self is the result of a negotiation of competing interpretations, rather than a fixed entity.
• Wheeler, J. A. (1990). Information, Physics, Quantum: The Search for Links.
— "It from Bit": Being as a result of the act of distinction. The chapter uses it as a confirmation that consciousness is an act of shaping reality through choice and focus.

███ References — Chapter 5: Time and the Hidden Field of Potential Matter
• Padmanabhan, T. (2005). Emergent Gravity and Dark Energy.
— Considers dark energy as a consequence of fundamental information processes. Supports the hypothesis of a hidden field of coordination that influences the structure of existence.
• Greene, B. (1999). The Elegant Universe.
— String Theory and the Multidimensional Universe: Oscillations as the Basis of Matter. Used in the chapter as an image of mutual oscillations and connections.
• Planck Collaboration.
— Cosmological observations of the acceleration of the expansion of the Universe and the distribution of dark energy. Mentioned as the empirical basis for the invisible field hypothesis.
• Shannon, C. (1948). A Mathematical Theory of Communication.
— Entropy as a measure of uncertainty. Supports the idea of communication as a mechanism for managing chaos and distinguishability.
• Schrödinger, E. (1944). What is Life?
— The concept of negative entropy as a vital principle. Used in the text to explain the role of consciousness as a structure that reduces chaos.
• Hawking, S. (2005). Information Loss in Black Holes.
— The assumption of information preservation at the event horizon. Cited to support the idea that the structure of distinction is preserved even when matter disappears.
• Wheeler, J. A. (1990). Information, Physics, Quantum: The Search for Links.
— Formula It from Bit: physical reality emerges from acts of distinction. Directly mentioned as a justification for the primacy of information over matter.
• Higgs, P. W. (1964). Broken Symmetries and the Masses of Gauge Bosons.
— The Higgs field as an example of how interaction with a field gives particles mass. Used as an analogy for Communicational Energy.
• QCD / Glueball Hypothesis.
— Modern quantum chromodynamics: 98% of the proton mass is formed by gluon fields - binding energy. Directly described in the chapter as an example of "binding creates mass".
• Banach, S. & Tarski, A. (1924). Sur la décomposition des ensembles de points.
— The paradox of restructuring objects without adding substance. Mentioned in the chapter as a mathematical analogy of "reassembling form through relationships."
• Lamb, W. E. & Retherford, R. C. (1947). Fine Structure of the Hydrogen Atom by a Microwave Method.

— Lamb shift as evidence of vacuum activity. Mentioned to illustrate how the "emptiness" contains fluctuations and interactions.
• **Casimir, H. B. G. (1948). On the Attraction Between Two Perfectly Conducting Plates.**
— Casimir effect: vacuum has power. Used as experimental proof of the existence of energy even in "nothingness".

References — Chapter 6: Man on the Path of Evolution
• **Darwin, C. (1859). On the Origin of Species.**
— Evolution as selection, acting not only at the level of organisms, but also at the level of their ability to communicate and coordinate. The chapter quotes directly and expands on the idea of communication as a factor in survival.
• **Turchin, P. (2006). War and Peace and War: The Rise and Fall of Empires.**
— The concept of "asabiya" (internal solidarity) as a key factor in the sustainability of communities is mentioned directly in the text of the chapter and discussed as an energetic basis for coordination in a group.
• **Wilson, E. O. (1975). Sociobiology: The New Synthesis.**
— Human values and altruism as a result of the evolutionary strategy of cooperation. The chapter appeals to the idea that the ability to interact could have been naturally selected.
• **Axelrod, R. & Hamilton, W. D. (1981). The Evolution of Cooperation.**
— Computer modeling of cooperation strategies. The text describes that mutual trust between agents can evolutionarily win in the long term.
• **Harari, Y. N. (2014). Sapiens: A Brief History of Humankind.**
— The idea of the evolution of Homo sapiens as a capacity for large-scale abstract communication. It is cited in the text as an example of the emergence of myths, rituals, and common symbols as a means of expanding cooperation.
• **Tomasello, M. (2008). Origins of Human Communication.**
— Communication as jointly directed attention and coordination of actions. The chapter mentions the idea of children sharing attention with their parents as the basis for the formation of language.
• **Dunbar, R. I. M. (1996). Grooming, Gossip, and the Evolution of Language.**
— A hypothesis about the origin of language as a means of social bonding of groups. The text directly uses the metaphor "gossip as the glue of society".
• **Varela, F. J., Thompson, E., & Rosch, E. (1991). The Embodied Mind.**
— Consciousness as an embodied process, embedded in environment and action. The chapter emphasizes that evolution is not a linear ladder, but a network that depends on the capacity for connectivity.
• **Bateson, G. (1972). Steps to an Ecology of Mind.**
— Ecosystem thinking and the idea of information as "differences that matter." The chapter notes that the human mind is not an autonomous process, but an element of a coherent network.

References — Chapter 7: Neurobiology of Communication
• **Hebb, D. O. (1949). The Organization of Behavior.**
— The principle of "neurons that fire together, wire together" (Hebbian learning) is mentioned in the text as the basis for the formation of stable connection patterns in the brain.
• **Friston, K. (2010). The Free-Energy Principle.**
—The brain as a system that minimizes predictive error. The chapter directly examines it as a mechanism for coordinating perception and action through internal models.
• **Damasio, A. (1999). The Feeling of What Happens.**
— Emotions as the basis for decision-making and coordination between body and mind.

The chapter emphasizes that communication involves an affective response built into the body.

• **Tononi, G. (2004). An Information Integration Theory of Consciousness.**
— Consciousness as a result of information integration in a system. The chapter is based on the idea that coordinated neural activity generates a sense of unity of experience.

• **Gazzaniga, M. S. (2005). The Ethical Brain.**
— The Brain as a Modular System with a Post-Factum Interpreter. The chapter touches on the idea that we "complete" the meaning after the impulse.

• **Rizzolatti, G., & Sinigaglia, C. (2008). Mirrors in the Brain.**
— Mirror neurons as the basis of empathy and understanding of others. The chapter directly states that "we seem to feel another" not through thinking, but through reflection.

• **LeDoux, J. (1996). The Emotional Brain.**
— Neural Circuits of Fear and Survival. The text talks about the immediate activation of ancient brain structures upon recognition of a threat - before comprehension.

• **Porges, S. W. (2011). The Polyvagal Theory.**
— The Role of the Vagus Nerve in Communication Responses: Defense, Withdrawal, Openness. The text describes how the biology of the body determines communicative behavior.

• **Barrett, L. F. (2017). How Emotions Are Made.**
— Emotions as predictive constructs, not universal reactions. The text emphasizes that emotional response is associated with modeling the future, and not just a reaction to the past.

• **Ramachandran, V. S. (2011). The Tell-Tale Brain.**
— The connection between the brain, body, and cultural patterns. The chapter mentions the idea that neural patterns are formed in response to social contexts and symbols.

References — Chapter 8: Games of Consciousness and the Movement of Matter

• **Bohr, N. (1928). The Quantum Postulate and the Recent Development of Atomic Theory.**
— The idea of complementarity: you can't observe a quantum system without interfering with it. The text of the chapter uses it as a basis for thinking about the relationship between the observer and reality.

• **Heisenberg, W. (1927). Über den anschaulichen Inhalt der quantentheoretischen Kinematik und Mechanik.**
— Uncertainty principle: it is impossible to measure both the position and momentum of a particle at the same time. Mentioned in the chapter as an illustration of the extreme dependence of knowledge on the act of observation itself.

• **Wigner, E. P. (1961). Remarks on the Mind-Body Question.**
— Consciousness as a necessary condition for the collapse of the wave function. The chapter directly touches on the idea: observation is not a neutral act, but a co-creation of reality.

• **Neumann, J. von (1955). Mathematical Foundations of Quantum Mechanics.**
— Mathematical formalization of the participation of consciousness in quantum measurement. The chapter refers to this idea as a possible explanation of the "co-influence" on matter.

• **Penrose, R. (1994). Shadows of the Mind.**
— An attempt to combine the theory of consciousness and quantum physics. The text uses it as a confirmation of the hypothesis about the deep connection between subjectivity and physical processes.

• **Kauffman, S. A. (1995). At Home in the Universe.**
— Autocatalytic systems and spontaneous self-organizing behavior. The text discusses as the basis of the logic of coordinating elements without an external center.
• **Barabási, A.-L. (2002). Linked: The New Science of Networks.**
— Laws of network formation and the emergence of attraction nodes. The chapter uses them to explain how individual "solutions" shape the logic of systems.
• **Tegmark, M. (2017). Life 3.0.**
— The idea of consciousness as a computing system that influences the trajectories of matter. Mentioned in the text when discussing the possibility of a self-learning universe.
• **Prigogine, I. (1980). From Being to Becoming.**
— Time as an irremovable factor of instability and creativity. The text discusses the transition from the "givenness" of matter to the "play" of changes and fluctuations.
• **Varela, F. J. (1997). Ethical Know-How: Action, Wisdom, and Cognition.**
— Awareness as a form of attunement with the world field. The chapter touches upon the topic of movement not as displacement, but as an act of presence in the flow.

References — Chapter 9: Reality Code. From Zero to One
• **Leibniz, G. W. (1703). Explication de l'Arithmétique Binaire.**
— Development of a binary system: 0 and 1 as a universal code. In the chapter it is cited directly as the basis of the digital description of reality and the connection of numbers with the logic of the universe.
• **Shannon, C. (1948). A Mathematical Theory of Communication.**
— Fundamentals of information theory: bit as the minimal unit of distinction. Mentioned in the chapter as the central basis for the transition from distinction to code and information structure.
• **Boole, G. (1854). An Investigation of the Laws of Thought.**
— Boolean algebra as a formalization of logical operations through 0 and 1. Used in the chapter as a logical-mathematical basis for digital processing of meaning.
• **Wheeler, J. A. (1990). Information, Physics, Quantum: The Search for Links.**
— The concept of *It from Bit*: physical reality as a result of the act of distinction. In the chapter, it occupies a central place in the transition from philosophy to digital ontology.
• **Turing, A. (1936). On Computable Numbers.**
— The idea of a universal machine and the generation of computable reality. Mentioned as the logical source of the possibility of constructing entire worlds from the simplest operations.
• **Dehaene, S. (2014). Consciousness and the Brain.**
— A study of how a neural network encodes meaning. In the text, it is used in the context of digital models of the brain and the recoding of signals into stable patterns.
• **Dennett, D. C. (1991). Consciousness Explained.**
— Consciousness as a virtual machine running on the code of differences. The chapter presents the idea that thinking does not require materiality, but relies on computable processes.
• **Tegmark, M. (2014). Our Mathematical Universe.**
— The Universe as a Mathematical Structure. The chapter supports the idea of code as a universal basis that unites information, logic, and physics.
• **Chaitin, G. (2005). Meta Math! The Quest for Omega.**
— The idea of mathematical randomness and the limitations of formal knowledge. The text touches upon this as the ultimate condition of the code Universe: structure is born on the edge of chaos.
• **Varela, F. J., Thompson, E., & Rosch, E. (1991). The Embodied Mind.**
— Embodied Knowledge: Code is not an abstraction, but part of living perception. The text discusses how zero and one acquire meaning only in the field of presence and action.

▦ References — Chapter 10: Digital Alchemy (2–5)

• **Shannon, C. (1948). A Mathematical Theory of Communication.**
— Communication as a distinction between possible states. The chapter confirms the transition from 2 (distinction) to 3 (context, order) as the logic of increasing structure.

• **Bateson, G. (1972). Steps to an Ecology of Mind.**
— The classic definition of information is "a difference that matters." The text uses it as the basis for the idea: 2 is not just a binary, but an act of meaningful selection.

• **Peirce, C. S. (1903). Pragmatism and Pragmaticism.**
— Triadic model of the sign: sign, object and interpretant. The text mentions the structure of "triplicity" as a mechanism for the emergence of meaning.

• **Jakobson, R. (1960). Linguistics and Poetics.**
— Functions of language: distinction, contextualization, hierarchy. The chapter refers to the construction of a semantic structure from the simplest differences (2 → 3 → 4).

• **Luhmann, N. (1984). Soziale Systeme.**
— Self-referential systems emerging through internal differences. The chapter plays on this as an example of the emergence of structure through observation of differences within the system itself.

• **Hofstadter, D. (1979). Gödel, Escher, Bach.**
— Ideas of recursion, self-reference and multi-layered semantic organization. Mentioned in the transition from 3 to 4 - as a stage of formation of stable patterns and "game of self-recognition".

• **Barabási, A.-L. (2002). Linked: The New Science of Networks.**
— Patterns of network growth and the emergence of "nodes" with increased significance. In the chapter appears as confirmation - when moving from 4 to 5, a non-linear structure and scalability arises.

• **Kauffman, S. A. (1993). The Origins of Order.**
— Self-organization in biological and information systems. Used as a model: from simple interactions (2) through stable triads (3), to a multitude of possible assemblies (4–5).

• **Deleuze, G., & Guattari, F. (1980). A Thousand Plateaus.**
— The concept of the rhizome: non-linear, non-hierarchical networks of differences. Mentioned as an alternative way of organizing structure - not through rigid logic, but through "assemblages without a center."

▦ References — Chapter 11: From 6 to 10. Algorithms Leading to Transcendence

• **Maslow, A. H. (1943). A Theory of Human Motivation.**
— Hierarchy of Needs. Mentioned in the chapter in the context of the transition from satisfying basic needs (levels 1–5) to going beyond—to meaning, integrity, transcendence.

• **Graves, C. W. (1970). Levels of Human Existence.**
— Hierarchy of Needs. Mentioned in the chapter in the context of the transition from satisfying basic needs (levels 1–5) to going beyond—to meaning, integrity, transcendence.

• **Beck, D. E., & Cowan, C. (1996). Spiral Dynamics.**
— Development of values and consciousness according to the spiral model. The text describes as a form of evolutionary communication - the transition from individual to integral thinking.

• **Wilber, K. (2000). Integral Psychology.**
— Multilevel development of consciousness: body, mind, spirit. The text mentions it as an attempt to synthesize psychological, spiritual and cultural growth algorithms.

• **Gebser, J. (1949). The Ever-Present Origin.**
— Structures of consciousness: archaic, magical, mythical, mental, integral. The chapter uses the logic of changing modes of perception as part of the movement through levels.
• **Morin, E. (1999). La Méthode.**
— Complexity, self-reflection, transition from linearity to nonlinear thinking. The chapter uses it as a basis for understanding "level 10" as a consciousness capable of holding contradictions and multitudes.
• **Teilhard de Chardin, P. (1955). The Phenomenon of Man.**
— The idea of the "Omega point" - the evolutionary goal of consciousness and humanity. The chapter mentions it as an image of the "vector of gathering" and spiritual coordination.
• **Turing, A. (1950). Computing Machinery and Intelligence.**
— Question: Can a machine think? The text poses this question as a parallel between growth algorithms and self-developing programs.
• **Goertzel, B. (2009). The Hidden Pattern.**
— The idea of consciousness as a self-organizing information structure. Used in the chapter when discussing how higher levels of consciousness operate as complex discrimination systems.

🔲 **References — Chapter 12. The Ladder of Communication Growth**
• **Maslow, A. H. (1968). Toward a Psychology of Being.**
— The transition from needs to self-realization and "peak experiences." The text uses it as a basis for the first level of communication growth — "getting out of deficit."
• **Graves, C. W. (1970). Levels of Human Existence.**
— The concept of ascending levels of adaptation of consciousness. The chapter mentions the "communication ladder" model as an inspiring source.
• **Vygotsky, L. S. (1934). Thought and Language.**
— Zone of proximal development: learning as a social process. The text uses it as an analogy - communication development occurs through joint activities.
• **Piaget, J. (1952). The Origins of Intelligence in Children.**
— Stages of cognitive development: from concrete to abstract. The chapter uses the idea that growth is possible through the expansion of thinking structures.
• **Bateson, G. (1972). Steps to an Ecology of Mind.**
— Logical levels of learning and change. The chapter mentions the idea of "second-order learning" as one of the levels of communication development.
• **Wilber, K. (2000). A Theory of Everything.**
— Integral model of development: consciousness, culture, body. The chapter reflects as a structure of multi-level growth of interaction.
• **Goleman, D. (1995). Emotional Intelligence.**
— Empathy, self-regulation and social skills as the basis for effective communication. The text notes this as the "level of conscious communication."
• **Bloom, B. S. (1956). Taxonomy of Educational Objectives.**
— Hierarchy of cognitive skills: from knowledge to synthesis. Used in the chapter as a model for the step-by-step development of depth of understanding in communication.
• **Csikszentmihalyi, M. (1990). Flow: The Psychology of Optimal Experience.**
— Flow as a state of maximum inclusion and coordination. Reflected in the chapter in the description of "higher levels of coordination".
• **Scharmer, O. (2009). Theory U: Leading from the Future.**
— Transformation through "listening to the future." The chapter explicitly mentions the U-model as a way to move from reactive to creative communication.

🔲 **References — Chapter 13: The Energy of Communication in Social Life**

• **Durkheim, É. (1912). Les formes élémentaires de la vie religieuse.**
— Collective rituals as a basis for social solidarity. In the chapter it is mentioned as a mechanism structuring a common space through coordinated actions.
• **Weber, M. (1922). Economy and Society.**
— The concept of the legitimacy of power through symbolic forms of communication. The text reflects the understanding of power as a derivative of consent and shared meaning.
• **Habermas, J. (1981). The Theory of Communicative Action.**
— Rationality that emerges in the process of dialogue. The chapter uses this as a rationale for constructive communication as an alternative to systemic violence.
• **Luhmann, N. (1984). Soziale Systeme.**
— Communication as a self-reproducing mechanism of social systems. The text uses the idea that society does not consist of people, but of communications between them.
• **Castells, M. (1996). The Rise of the Network Society.**
— Networks as the basis of modern economy and power. The chapter uses as evidence: social energy is formed in the flows of communication, not in institutions.
• **Bourdieu, P. (1977). Outline of a Theory of Practice.**
— The concept of habitus and symbolic capital. The text notes that speech, gestures, and style of interaction are a form of power and social positioning.
• **Pentland, A. (2014). Social Physics.**
— Collective behavior as a result of the spread of signals in a group. The text refers to it as confirmation: agreement creates trust and efficiency.
• **Putnam, R. D. (2000). Bowling Alone.**
— Social capital and its decline in the context of individualization. The chapter discusses how the destruction of connections and the loss of Communicational Energy in society causes.
• **Rogers, E. M. (1962). Diffusion of Innovations.**
— Spreading ideas through social networks. The text reflects: new meanings are not consolidated through force, but through acceptance within the framework of online transmission.
• **Morin, E. (1999). La Méthode.**
— Complexity as a condition of living systems. The chapter supports the idea that society is not a hierarchy, but a field of agreed differences.

⬛ References — Chapter 14: Language as a Basis for the Formation of Society
• **Saussure, F. de (1916). Cours de linguistique générale.**
— The structure of language as a system of differences between signs. In the chapter it is mentioned as the basis of the idea: meaning is created not in words, but in the relationships between them.
• **Jakobson, R. (1960). Linguistics and Poetics.**
— Functions of language: expressive, referential, metalinguistic, etc. The chapter uses this to demonstrate that language is not just a code, but a multi-layered mechanism of action.
• **Peirce, C. S. (1903). Pragmatism and Pragmaticism.**
— Triadic model of the sign: sign, object and interpretant. In the chapter it is mentioned as the logic of generating meaning through relationships, and not through unambiguous meanings.
• **Lotman, Y. M. (1984). The Semiotics of Culture.**
— Culture as a secondary model of reality generated by linguistic structures. The chapter supports the idea that language shapes not just communication, but a worldview.
• **Eco, U. (1976). A Theory of Semiotics.**
— Language as a system of codes that transmit not only information but also values. The chapter examines how each language creates its own version of the world.

• **Wittgenstein, L. (1953). Philosophical Investigations.**
— Language as a form of life, meaning in use. The chapter draws a line: words live in context and interaction, not in a dictionary.
• **Chomsky, N. (1957). Syntactic Structures.**
— Universal grammar as an innate structure of language. Mentioned in the chapter in the context of the discussion of the connection between language, thought, and innate patterns of understanding.
• **Lakoff, G., & Johnson, M. (1980). Metaphors We Live By.**
— Metaphor as the basis of conceptual thinking. The chapter emphasizes that metaphors do not decorate speech - they shape the perception of reality.
• **Austin, J. L. (1962). How to Do Things with Words.**
— Speech Act Theory: Language as Action. The chapter uses the idea that speech does not simply communicate, but changes reality.
• **Barthes, R. (1957). Mythologies.**
— Language as a way of mythologizing the everyday. The chapter uses it to explain how language can create the illusion of "naturalness" of meanings.

References — Chapter 15: Faith as a Universal Communication Protocol

• **James, W. (1902). The Varieties of Religious Experience.**
— Faith as a personal experience of the transcendent, not reducible to institutional religion. The chapter mentions it as the basis for experiencing a deep connection - beyond rituals and doctrines.
• **Eliade, M. (1957). The Sacred and the Profane.**
— Separation of sacred and profane space. The chapter uses the idea that faith creates "knots of meaning" in everyday reality.
• **Durkheim, É. (1912). Les formes élémentaires de la vie religieuse.**
— Religion as a form of social connection and solidarity. The chapter uses as evidence: faith is a protocol for coordinating community.
• **Barthes, R. (1957). Mythologies.**
— Modern myths as codes of belief embedded in language. The text emphasizes that even in secular societies, "beliefs" embedded in culture continue to operate.
• **Otto, R. (1917). Das Heilige.**
— The concept of the numinous: the experience of the sacred as both terrifying and attractive. The chapter touches on the basis of the sense of "connection with the beyond."
• **Jung, C. G. (1959). Psychology and Religion.**
— The Archetype of Faith as an Innate Structure of the Psyche. The chapter uses the idea that the image of God or the Absolute is an internal form of connection with the whole.
• **Varela, F. J., Thompson, E., & Rosch, E. (1991). The Embodied Mind.**
— Perceiving the world as a connection and attunement, not as distant knowledge. The chapter refers to faith as a way of "being in touch."
• **Buber, M. (1923). Ich und Du.**
— Faith as a dialogical relationship with the Absolute, not objective knowledge. The chapter mentions it as a form of internal relationship that creates a "field of presence".
• **Eckhart, M. (14th century). Sermons.**
— The image of "God within" and silence as a form of higher connection. In the text of the chapter it is used as an example of experience without intermediaries, outside the linguistic form.
• **Capra, F. (1975). The Tao of Physics.**
— Crossroads between physics and mystical experience. The chapter mentions the possibility of reconciling science and faith as forms of knowledge of the connection.

■ References — Chapter 16: Digital Institutions and Future Communication
• **Castells, M. (1996). The Rise of the Network Society.**
— Society as a Network: Power, Economy, and Culture in the Context of Network Logic. The chapter uses this as a basis for discussing digital institutions as supranational structures.
• **Beniger, J. R. (1986). The Control Revolution.**
— The emergence of the information society as a response to the crisis of governance. The chapter mentions as a confirmation of the idea: digital institutions are a new form of coordination of complex systems.
• **Lessig, L. (1999). Code and Other Laws of Cyberspace.**
— Program code as a new form of "law". The chapter uses the idea that the architecture of the digital environment defines the boundaries and rules of behavior.
• **Zuboff, S. (2019). The Age of Surveillance Capitalism.**
— The Surveillance Economy and the Power of Algorithms. The chapter refers to the fact that digital institutions often govern implicitly — through the architecture of behavior, rather than through declared norms.
• **Morozov, E. (2011). The Net Delusion.**
— The illusion of digital freedom and the risks of technological control. The text discusses how the digital environment is not neutral, but carries a hidden ideology.
• **Ostrom, E. (1990). Governing the Commons.**
— Self-organizing governance of common resources. The chapter uses it as a model of a possible digital community based on transparency and trust.
• **Berners-Lee, T. (1999). Weaving the Web.**
— Principles of Decentralization and Open Standards. The chapter refers to the original concept of the Internet as a field of equal connections - before the advent of platform centralization.
• **Pentland, A. (2014). Social Physics.**
— Digital traces and patterns of collective behavior. The chapter supports the idea that the institutions of the future will be built not on formal rules, but on the analysis of real communication flows.
• **Tapscott, D. & Tapscott, A. (2016). Blockchain Revolution.**
— Blockchain as an architecture of trust without intermediaries. The text mentions it as an example of an institution based on transparent communication rather than hierarchy.
• **Lanier, J. (2010). You Are Not a Gadget.**
— Critique of digital reductionism. The chapter discusses the danger that humans cease to be subjects and become functions of the digital system.

■ References — Chapter 17: Communication Literacy
• **Freire, P. (1970). Pedagogy of the Oppressed.**
— Critical literacy as a tool for liberation. The chapter uses this as a basis: a person becomes a subject when he/she becomes aware of the structure of the communication in which he/she participates.
• **Postman, N. (1985). Amusing Ourselves to Death.**
— The decline of critical thinking in the age of entertainment. The chapter mentions as a warning: without literacy, communication turns into a stream of manipulation.
• **McLuhan, M. (1964). Understanding Media.**
— "Media is the message": form influences content. The chapter uses as a basic idea: literacy is not only the ability to read, but also the ability to read the *environment*.
• **Orwell, G. (1949). 1984.**
— Language as a tool for controlling and resetting thinking. The chapter discusses how, without awareness, language can become a mechanism for losing meaning.

- **Lakoff, G. (2004). Don't Think of an Elephant!**
— Frames as the basis of political and value communication. The chapter uses this as confirmation: literacy is the ability to see how it is said, and not just *what*.
- **Kahneman, D. (2011). Thinking, Fast and Slow.**
— Systems of Thought and Cognitive Biases. The chapter emphasizes that conscious communication requires effort — otherwise, automatism and predictability kick in.
- **Rosenberg, M. (2003). Nonviolent Communication.**
— Language as a means of connection, not pressure. The chapter examines it as a practice that forms literacy in emotional and semantic transmission.
- **Habermas, J. (1981). The Theory of Communicative Action.**
— Rationality that emerges from open dialogue. The text supports the idea that true communication literacy requires the ability to dialogue, not argue.
- **Bohm, D. (1996). On Dialogue.**
— Dialogue as a process of joint thinking. The chapter uses as an ideal: communication in which a position is not asserted, but a common field is explored.
- **Turkle, S. (2015). Reclaiming Conversation.**
— The loss of deep speech in the digital world. The text mentions it as a symptom: we need to relearn how to hear and speak – despite the abundance of words.

References — Chapter 18: Consciousness and Beauty: Creation and Perception

- **Plato. (ca. 380 BCE). Symposium / Phaedrus.**
— Beauty as a manifestation of higher ideas. In the chapter it is mentioned as a philosophical basis: the perception of beauty is a form of touching the truth.
- **Kant, I. (1790). Critique of Judgment.**
— Aesthetic judgment as an act of free harmony between imagination and reason. The text uses the idea that beauty is born in the harmony of the felt and the thought.
- **Scruton, R. (2009). Beauty.**
— Beauty as a form of meaning and order that evokes response. The chapter supports the idea that aesthetics is not a subjective decoration but a profound response to coherence.
- **Berleant, A. (1991). Art and Engagement.**
— Aesthetics as an included, bodily perception. The chapter draws a line: beauty requires not distance, but participation and attunement.
- **Ramachandran, V. S., & Hirstein, W. (1999). The Science of Art.**
— Neuroaesthetics: Principles of Perception of Visual Harmony. The text mentions as a confirmation that beauty operates through rhythm, symmetry, contrast and surprise.
- **Zeki, S. (1999). Inner Vision: An Exploration of Art and the Brain.**
— Aesthetic perception as a function of specialized brain areas. The chapter uses the idea that the brain responds to beauty as a coherent complexity.
- **Damasio, A. (1999). The Feeling of What Happens.**
— Aesthetic experience as a combination of feeling, image and meaning. The chapter supports the idea that beauty is inseparable from bodily "living".
- **Whitehead, A. N. (1929). Process and Reality.**
— Aesthetic experience as a combination of feeling, image and meaning. The chapter supports the idea that beauty is inseparable from bodily "living".
- **Varela, F. J., Thompson, E., & Rosch, E. (1991). The Embodied Mind.**
— Perception as active, bodily participation. The text of the chapter emphasizes: beauty is not an object, but a field of attunement between the subject and the world.
- **Nietzsche, F. (1872). The Birth of Tragedy.**
— Beauty as a tension between Apollonian order and Dionysian chaos. The chapter touches on the idea that living creativity is born on the edge between form and decay.

■ References — Chapter 19: Positive as the Main Impulse of Being
• **Frankl, V. E. (1946). Man's Search for Meaning.**
— Meaning as an internal support capable of overcoming suffering. In the chapter it is used as proof that even in extreme conditions a person can choose the side of life.
• **Maslow, A. H. (1968). Toward a Psychology of Being.**
—Positivity not as naivety, but as a manifestation of mature, self-realized consciousness. The chapter cites it in the context of "growth motivation" - movement not from fear, but towards meaning.
• **Fredrickson, B. L. (2001). The Role of Positive Emotions in Positive Psychology.**
— Broaden and Build Theory: Positive Emotions Increase Adaptive Resources. The chapter refers to the scientific basis for positive energy as an evolutionary asset.
• **Seligman, M. E. P. (2011). Flourish.**
— The Psychology of Prosperity: Positivity as a Structural Part of Sustainable Living. The chapter uses the idea that positivity is not an emotion, but a strategic attitude towards being.
• **Csikszentmihalyi, M. (1990). Flow: The Psychology of Optimal Experience.**
— Flow as a form of positive presence in activity. In the chapter it is mentioned as a state in which the duality between the subject and the task disappears.
• **David, S. (2016). Emotional Agility.**
— Positivity not as repression, but as flexibility in experiencing all states. The chapter emphasizes that positivity is not in avoiding pain, but in the ability to integrate it.
• **Nietzsche, F. (1882). The Gay Science.**
— Amor fati — "love of fate": accepting life in its entirety. The chapter mentions a paradoxically masculine form of positivity — affirming meaning even in the absurd.
• **Damasio, A. (2003). Looking for Spinoza.**
— Emotions as the basis of reason and action. The chapter supports the idea that positivity is not an adornment of logic, but its source in bodily dynamics.
• **Teilhard de Chardin, P. (1955). The Phenomenon of Man.**
— Evolution as a movement towards ever greater coordination and integrity. In the chapter, the positive is interpreted as the power of gathering and increasing the coherence of being.

■ References — Chapter 2.0: Contract with Life: Perpetual Motion and the Great Observer
• **Wheeler, J. A. (1990). Information, Physics, Quantum: The Search for Links.**
— The *It from Bit* concept: reality emerges from acts of observation. The chapter uses the central idea: the observer does not record - he creates.
• **Neumann, J. von (1955). The Mathematical Foundations of Quantum Mechanics.**
— Consciousness as a necessary element of the collapse of the wave function. In the chapter, this is the key to understanding: life is participation, not external analysis.
• **Whitehead, A. N. (1929). Process and Reality.**
— The universe as a flow of events, where each is an act of perception. The chapter uses as a basis the idea: to be means to enter into an act of consonance.
• **Spinoza, B. (1677). Ethics.**
— God or nature as a single field of being. The chapter refers to the idea that everything is an expression of one essence - and the contract with life is an attunement with this integrity.
• **Teilhard de Chardin, P. (1955). The Phenomenon of Man.**
— Movement to the Omega point: consciousness as an evolutionary vector. In the chapter it is used as an image of a large contract with Being: participation in gathering.

• **Kauffman, S. A. (1993). The Origins of Order.**
— Self-organization and emergence as an internal impulse of life. The chapter confirms the idea: life is not a chain, but a flashing pattern of connection.
• **Barad, K. (2007). Meeting the Universe Halfway.**
— Agent reality: the observer and the world co-create the event. The chapter directly supports the idea: a contract with life is not an act of will, but an act of joint attunement.
• **Bateson, G. (1972). Steps to an Ecology of Mind.**
— The mind is not in the head, but in patterns of connection. The chapter consolidates as a conclusion: to live means to distinguish, to connect, to be part of a great dialogue.
• **Nagarjuna (ca. II c.). Mūlamadhyamakakārikā.**
— Emptiness as the interdependence of all things. The text mentions it as a philosophical basis for the rejection of rigid subjectivity and the transition to an open observer.
• **Bohr, N. (1934). Atomic Theory and the Description of Nature.**
— The principle of complementarity: the inseparability of subject and object. The chapter confirms the idea: to observe means to participate.

GLOSSARY OF TERMS

Core concepts in the framework of the Communicational Energy Hypothesis

Act of Communication
The process of establishing a connection between elements of a system, leading to the emergence of structure, action, or new meaning. **Anti-Entropy**
A force that counters disorder. Describes the tendency of conscious systems to organize themselves through interaction and meaningful exchange.

Assemblage Point
A moment of convergence between inner model and outer signal—where insight or transformation emerges.

Association / Interconnectedness
An ontological principle that nothing exists in isolation—everything is entangled in networks of feedback, influence, and mutual dependence.

Attunement
The ability of systems to enter alignment—emotionally, cognitively, or structurally—through shared rhythm or logic.

Cognitive Module
Practical exercises in the book designed to develop awareness, reflection, and deeper connection through focused attention.

Communication
More than information exchange—an ontological mechanism by which reality arises through alignment and structured interaction.

Communicational Energy
A central hypothesis: not a metaphor but a real force enabling connection, structure, order, and awareness—like gravity for meaning.

Consciousness
Not a mirror but a constructor of reality. The active mechanism behind perception, prediction, and meaning formation.

Consciousness as Law
The metaphysical idea that consciousness is embedded in the fabric of the universe, sustaining order and developmental direction.

Construction of Meaning
The process by which perception and information crystallize into coherent models of understanding and direction.

Dark Matter / Dark Energy
Unobservable forces that shape cosmic structure. In the book, it represents latent fields of unrealized form and connection.

Digit as Symbol
Numbers not as quantities, but as archetypes. 0–10 represents the full cycle of evolution

from void to integrated totality.

Digital Alchemy
The notion that numbers carry quantity and archetypal meaning. Each number from 0 to 10 marks a stage in the evolution of form and consciousness.

Entropy
A measure of disorder and loss of structure. In the book, a counterforce to consciousness and connection—associated with disintegration and meaning-loss.

Formula of Action
The expression of communicational energy as directed change—decision, creation, movement.

Fractal
A pattern where the part mirrors the whole. Represents the recursive nature of meaning and structure through communication.

Global Workspace Theory (GWT)
A cognitive model of consciousness describing how attention binds various neural processes into a coherent, unified system.

Gluons / Glueballs
Particles that bind quarks together. Symbolic in the book as an illustration that mass arises from interaction, not from the isolated object itself.

Holographic Principle
The idea that all the information of the universe may be encoded on a two-dimensional surface. Used here as an analogy for the informational field of communication.

Information Field
An invisible medium where energy, matter, and consciousness are unified through exchange, response, and the structuring of meaning.

Inner Narrative
A form of self-awareness in which a person structures experience, action, and meaning through internal dialogue.

Kabbalah
A mystical Jewish tradition interpreting creation as the descent of divine light through ten levels (Sefirot). Used here as a metaphor for the communication algorithm from potential to form.

Matter
Not a static substance, but a manifestation of relational coherence. What is connected becomes tangible.

Meaning as Structure
Meaning is not content, but a form arising from interaction and differentiation—where focused connection gives rise to significance.

Mirror Neurons
Neurons that fire both when an action is performed and when it is observed in others. Underlying mechanism for empathy and attunement.

Neural Networks
Artificial or biological systems capable of learning, processing, and forming associations. In the book, a model of distributed intelligence.

Neural Synchronization
Alignment of brain rhythms between individuals—underlying genuine understanding, empathy, and resonance.

Ordering
The process of shaping chaos into structure—achieved through meaningful interaction and connection.

Potential Field

A domain of unrealized possibilities. Associated with dark matter—a space where structure can emerge through communicational energy.

Predictive Coding
A theory that the brain continuously generates hypotheses about the world and adjusts them based on incoming sensory data to minimize prediction error.

Quantum
The smallest indivisible unit of energy or interaction. Symbolizes how even the tiniest acts of connection shape reality.

Quantum Entanglement
The phenomenon of instant interdependence between particles across distance—used as a metaphor for non-local, deep communication.

Quantum Mechanics
The branch of physics dealing with subatomic phenomena. In the book, it supports the view that interaction and information precede substance.

Resonance
A state of aligned interaction that amplifies connection, coherence, and systemic integrity.

Response
A structured continuation of connection—not merely reaction, but co-creation of coherence.

Sacred Geometry
The idea that numbers and forms encode universal laws of structure and meaning—mathematics as a language of reality.

Sefirot
The ten emanations in Kabbalistic cosmology. Used here as an archetypal path of unfolding consciousness and form.

String Theory
A physical theory suggesting that particles are vibrational strings. Mirrors the book's view of the universe as patterns of rhythm and relation.

Superposition
A state in which all possible forms coexist until observed or linked—used metaphorically for unrealized potential in systems.

Synergetics
The science of self-organizing systems. Employed here to explain how connection among elements generates emergent order.

Transformation of Information
The capacity to not only transmit data but to translate it into action, meaning, and systemic form.

www.ingramcontent.com/pod-product-compliance
Lightning Source LLC
Chambersburg PA
CBHW021216130626
46554CB00004B/1240